AF752215

TRANSPORTATION ISSUES, POLICIES AND R&D

# UNINTENDED OR SUDDEN ACCELERATION IN AUTOMOBILES

# TRANSPORTATION ISSUES, POLICIES AND R&D

Additional books in this series can be found on Nova's website under the Series tab.

Additional E-books in this series can be found on Nova's website under the E-book tab.

# SAFETY AND RISK IN SOCIETY

Additional books in this series can be found on Nova's website under the Series tab.

Additional E-books in this series can be found on Nova's website under the E-book tab.

TRANSPORTATION ISSUES, POLICIES AND R&D

# UNINTENDED OR SUDDEN ACCELERATION IN AUTOMOBILES

DENISE M. MCMANN
EDITOR

NOVINKA

Nova Science Publishers, Inc.
*New York*

Copyright © 2011 by Nova Science Publishers, Inc.

**All rights reserved.** No part of this book may be reproduced, stored in a retrieval system or transmitted in any form or by any means: electronic, electrostatic, magnetic, tape, mechanical photocopying, recording or otherwise without the written permission of the Publisher.

For permission to use material from this book please contact us:
Telephone 631-231-7269; Fax 631-231-8175
Web Site: http://www.novapublishers.com

**NOTICE TO THE READER**

The Publisher has taken reasonable care in the preparation of this book, but makes no expressed or implied warranty of any kind and assumes no responsibility for any errors or omissions. No liability is assumed for incidental or consequential damages in connection with or arising out of information contained in this book. The Publisher shall not be liable for any special, consequential, or exemplary damages resulting, in whole or in part, from the readers' use of, or reliance upon, this material. Any parts of this book based on government reports are so indicated and copyright is claimed for those parts to the extent applicable to compilations of such works.

Independent verification should be sought for any data, advice or recommendations contained in this book. In addition, no responsibility is assumed by the publisher for any injury and/or damage to persons or property arising from any methods, products, instructions, ideas or otherwise contained in this publication.

This publication is designed to provide accurate and authoritative information with regard to the subject matter covered herein. It is sold with the clear understanding that the Publisher is not engaged in rendering legal or any other professional services. If legal or any other expert assistance is required, the services of a competent person should be sought. FROM A DECLARATION OF PARTICIPANTS JOINTLY ADOPTED BY A COMMITTEE OF THE AMERICAN BAR ASSOCIATION AND A COMMITTEE OF PUBLISHERS.

Additional color graphics may be available in the e-book version of this book.

**LIBRARY OF CONGRESS CATALOGING-IN-PUBLICATION DATA**

Unintended or sudden acceleration in automobiles / editor, Denise M. McMann.
p. cm.
Includes index.
ISBN 978-1-61728-983-5 (softcover)
1. Automobiles--Fuel systems. 2. Automobiles--Defects. 3. Fault location (Engineering) 4. Automobiles--Defects--Reporting--Government policy--United States. I. McMann, Denise M.
TL214.F8U55 2010
363.12'5--dc22

2010026973

*Published by Nova Science Publishers, Inc. † New York*

# CONTENTS

# PREFACE

Congress is considering legislation to strengthen federal regulation of auto safety, in response to hundreds of reported accidents, and more than 50 fatalities, that may be linked to sudden acceleration in certain makes of Toyota and Lexus vehicles. Toyota, under pressure from the National Highway Traffic Safety Administration (NHTSA), has recalled more than 6 million autos in the United States, and for a time stopped production of certain new-model vehicles in an effort to address the problems, which is one of the largest such efforts in recent history. This book examines the problem of unintended or sudden acceleration of automobiles occurring during the past year and what mitigating challenges and procedures lay ahead for Congress and automakers.

Chapter 1 - Congress is considering legislation to strengthen federal regulation of auto safety, in response to hundreds of reported accidents, and more than 50 fatalities, that may be linked to sudden acceleration in certain makes of Toyota and Lexus vehicles. Toyota, under pressure from the National Highway Traffic Safety Administration (NHTSA), has recalled more than 6 million autos in the United States, and for a time stopped production of certain new-model vehicles in an effort to address the problems—one of the largest such efforts in recent history. But lawmakers and consumer advocates say federal oversight has been inadequate, given that NHTSA began investigating reports of sudden acceleration in certain Toyotas in 2002. Internal corporate documents indicate that Toyota was slow to address other vehicle quality concerns. NHTSA in April 2010 fined Toyota $1 6.375 million, the maximum allowed by law, for failing to quickly notify the U.S. Department of Transportation (DOT) of an accelerator pedal defect in some cars.

Adding to the complexity of the situation is continuing uncertainty about the cause of the Toyota incidents. Despite the recalls, a series of NHTSA investigations, and internal tests by Toyota engineers since 2002, there is not a consensus as to whether reported incidents of sudden acceleration have been caused by sticking accelerator pedals, poorly designed floor mats, driver error, glitches in the electronic throttle systems, or some combination of factors. DOT officials have expanded their investigation into Toyota operations, are taking a broad look at the increasingly complex electronics in all U.S. automobiles, and may recommend tougher standards for vehicles with electronic accelerators. At the same time, Toyota has announced its own stepped-up safety initiatives, including installing brake-override software (a system that gives priority to braking when the brake and accelerator are depressed at the same time) in some recalled and new-model cars, setting up an outside investigative panel headed by former Transportation Secretary Rodney Slater, and changing its corporate structure.

Lawmakers are considering a variety of issues as they craft a possible legislative response to the problems. Some Members of Congress are concerned that NHTSA did not conduct timely, thorough investigations of consumer complaints about Toyota vehicles and are examining whether the agency has sufficient staff and expertise, particularly in electrical and software engineering, to keep pace with the increasing complexity of autos. NHTSA has two electrical engineers on staff. Secretary of Transportation Ray LaHood has enlisted NASA engineers to assist NHTSA with its review.

Another issue is whether the 2000 TREAD Act, enacted in response to deaths and accidents linked to Firestone tires, has been effective. The act was designed to spur early warning of safety concerns, compel auto manufacturers to expeditiously address problems, and improve consumer information. The number of automotive recalls has increased since the TREAD Act was passed, but most of the recalls are voluntary, in part due to the lengthy legal process required in mandating a recall. Further, the DOT's $1 6.375 million fine against Toyota is based on the DOT's determination that the company did not comply with the TREAD Act requirements to quickly report defects. Congress is also examining whether Toyota problems are signs of broader issues in the automobile industry. Electronics and software in automobiles account for as much as 40% of the cost of producing a car, and the systems are advancing to the point that they are used to control more essential safety features. U.S. and foreign automakers carry out extensive testing of software and electronics based on industry guidelines and their own, internal standards. The quality control system is less rigorous than for other areas of the transportation sector, however, such as aviation.

Chapters 2 through 7 feature testimony before the U. S. Senate.

Chapter 8 - Historically, vehicle engine speed was mechanically controlled by use of a linkage or cable to the throttle valve assembly. The introduction of computerized engine controls has given vehicle engineers the added capability to electrically control fuel delivery, ignition, and most recently the throttle valve. By providing electrical throttle control through the means of an on-board vehicle computer, or Electronic Control Module (ECM), a number of engineering advantages were realized. Vehicle manufacturers were able to increase vehicle stability, achieve better fuel economy, reduce emissions, and eliminate components such as cruise control. One of the downside tradeoffs for these beneficial performance advantages, was increased electrical and electronic complexity. Vehicle manufacturers clearly recognized the important requirement for ETC systems to perform exactly as they intended. A failure of the electrical circuits, sensors, wiring, or actuators could potentially result in a runaway engine. Electronic Throttle Control (ETC) systems needed the added redundancy of certain sensors and electrical circuits to ensure safe and reliable operation. In addition, the ECM's were programmed to detect operational abnormalities or defects in ETC components and their related electrical circuits. The intent was to build an ETC system that would always "fail-safe" in the event of potential problem.

Vehicle engineers also needed to build in driver warning capability for ETC systems. In the event that a vital ETC component or electrical circuit defect has been detected, the Malfunction Indicator Lamp (MIL) is illuminated to inform the vehicle operator of a problem. At the same time, to aid in vehicle diagnostics and repair, a Diagnostic Trouble Code (DTC) will have been recorded in the ECM memory. Each possible DTC has an identifying alpha-numeric code to aid vehicle repair of computer associated components. The recorded DTC, along with a "freeze-frame" of captured vehicle data, could be retrieved later by technicians using a diagnostic scan tool. Depending on the seriousness of the ETC malfunction, the ECM will typically have been preprogrammed to operate the engine with limited RPM or substantially reduced power. In the event that mechanical components in the throttle body were to fail, the ECM has the added fail-safe capability to cut-off fuel delivery to limit engine speed. In the most serious cases of component failure, the ECM's fail-safe strategy could have the engine speed reduced to idle. This reduced power mode of operation would remain in the vehicle until proper repairs are made. All of these ETC fail-safe strategies were logically put in place to keep the vehicle safe and driver controllable at all times. However, for fail-safe vehicle operation, the ECM must have the programmed ability to detect a number of potential ETC malfunctions.

The ECM, in addition to operating the engine for optimum performance and emissions, is constantly monitoring sensors and circuits for incorrect or illogical values. Current production vehicles are required to meet standards for the second generation of On-Board Diagnostics (OBD II). An OBD II vehicle runs self-tests or monitors of the ECM circuitry that are typically classified as one or two trip faults. If a sensor or output circuit was to exceed the ECM's programmed threshold values, the computer should set a DTC. However, depending on the component being monitored, a MIL may not illuminate until the ECM sees the component fail a second time. Two trip fault detection logic is used for less definitive and less serious failures that are not easily distinguished as a problem area. One trip fault detection logic is used for more serious or readily identifiable problems. One trip fault monitors will set a DTC and turn on the MIL after only one instance of detection. Typically, one trip faults include electrical component, electrical circuit malfunctions, detected values that could cause engine damage, or potential safety concerns. Properly operating ETC system circuits and components are vital to safe vehicle operation. All DTC's for Toyota's ETC systems are identified in the service literature as "1 trip detection logic" with ECM code setting ability within seconds of circuit fault detection.

In: Unintended or Sudden Acceleration
Editor: Denise M. McMann
ISBN: 978-1-61728-983-5
© 2010 Nova Science Publishers, Inc.

*Chapter 1*

# UNINTENDED ACCELERATION IN PASSENGER VEHICLES

***Suzanne M. Kirchhoff and David Randall Peterman***

## SUMMARY

Congress is considering legislation to strengthen federal regulation of auto safety, in response to hundreds of reported accidents, and more than 50 fatalities, that may be linked to sudden acceleration in certain makes of Toyota and Lexus vehicles. Toyota, under pressure from the National Highway Traffic Safety Administration (NHTSA), has recalled more than 6 million autos in the United States, and for a time stopped production of certain new-model vehicles in an effort to address the problems—one of the largest such efforts in recent history. But lawmakers and consumer advocates say federal oversight has been inadequate, given that NHTSA began investigating reports of sudden acceleration in certain Toyotas in 2002. Internal corporate documents indicate that Toyota was slow to address other vehicle quality concerns. NHTSA in April 2010 fined Toyota $1 6.375 million, the maximum allowed by law, for failing to quickly notify the U.S. Department of Transportation (DOT) of an accelerator pedal defect in some cars.

Adding to the complexity of the situation is continuing uncertainty about the cause of the Toyota incidents. Despite the recalls, a series of NHTSA investigations, and internal tests by Toyota engineers since 2002, there is not a consensus as to whether reported incidents of sudden acceleration have been caused by sticking accelerator pedals, poorly designed floor mats, driver error, glitches in the electronic throttle systems, or some combination of factors. DOT officials have expanded their investigation into Toyota operations, are taking a broad look at the increasingly complex electronics in all U.S. automobiles, and may recommend tougher standards for vehicles with electronic accelerators. At the same time, Toyota has announced its own stepped-up safety initiatives, including installing brake-override software (a system that gives priority to braking when the brake and accelerator are depressed at the same time) in some recalled and new-model cars, setting up an outside investigative panel headed by former Transportation Secretary Rodney Slater, and changing its corporate structure.

Lawmakers are considering a variety of issues as they craft a possible legislative response to the problems. Some Members of Congress are concerned that NHTSA did not conduct timely, thorough investigations of consumer complaints about Toyota vehicles and are examining whether the agency has sufficient staff and expertise, particularly in electrical and software engineering, to keep pace with the increasing complexity of autos. NHTSA has two electrical engineers on staff. Secretary of Transportation Ray LaHood has enlisted NASA engineers to assist NHTSA with its review.

Another issue is whether the 2000 TREAD Act, enacted in response to deaths and accidents linked to Firestone tires, has been effective. The act was designed to spur early warning of safety concerns, compel auto manufacturers to expeditiously address problems, and improve consumer information. The number of automotive recalls has increased since the TREAD Act was passed, but most of the recalls are voluntary, in part due to the lengthy legal process required in mandating a recall. Further, the DOT's $16.375 million fine against Toyota is based on the DOT's determination that the company did not comply with the TREAD Act requirements to quickly report defects. Congress is also examining whether Toyota problems are signs of broader issues in the automobile industry. Electronics and software in automobiles account for as much as 40% of the cost of producing a car, and the systems are advancing to the point that they are used to control more essential safety features. U.S. and foreign automakers carry out extensive testing of software and electronics based on industry guidelines and their own, internal standards. The quality control system is less rigorous than for other areas of the transportation sector, however, such as aviation.

# INTRODUCTION

In August 2009, a California Highway Patrol officer was driving family members in a 2009 Lexus ES350 rental car when it accelerated to a high speed. The driver was not able to slow the car; it clipped another vehicle and careened off the road into a ravine, killing all four occupants. A passenger called 9-1-1 prior to the crash; the resulting recording captured the harrowing final seconds of the incident. The combination of the 9-1-1 audio and the driver's presumed driving expertise focused a great deal of attention on the accident and the broader issue of sudden unintended acceleration in Toyota vehicles.

The National Highway Traffic Safety Administration (NHTSA), an operating agency within the Department of Transportation (DOT) with chief responsibility for ensuring automobile safety, has received hundreds of consumer complaints about unintended acceleration in Toyotas going back at least to 2000. These complaints are not all the same; included within the general characterization of "sudden unintended acceleration" are instances where (1) a vehicle suddenly accelerates without any initiative by the driver, (2) a vehicle maintains its speed or continues to accelerate after a driver has released the accelerator pedal, and (3) a vehicle accelerates and the driver is unable to stop the vehicle by braking.

NHTSA has conducted eight examinations, including three investigations, of unexpected acceleration in Toyotas[1] during the past eight years. Toyota also has carried out its own internal studies. It was not until recently, however, that the NHTSA and Toyota investigations spurred major action. Toyota in 2007 issued a recall to address a body trim issue in Toyota Siennas that could interfere with the accelerator pedal. Since October 2009, Toyota has recalled more than 6 million Corolla, Camry, and Lexus ES model vehicles in the United States in an effort to remedy two defects identified as causing acceleration incidents: entrapment of the accelerator pedal by floor mats, and "sticking" accelerator pedals (i.e., a tendency for some accelerator pedals to return slowly after being released).[2] Toyota also temporarily stopped production of eight types of vehicles.[3] Still, the recalls have not resolved all the reported cases of unexpected acceleration. Further, some vehicle owners have complained of problems even after Toyota has reworked their accelerators as part of its recalls. Starting in February 2010, Congress began a series of hearings to look into the issue, Toyota's actions, and NHTSA's authority, resources, and response to this issue.

The probe has expanded beyond potential problems in Toyota vehicles. Secretary of Transportation Ray LaHood announced on March 30, 2010, that the DOT had asked the National Academy of Sciences (NAS) to investigate unintended acceleration and electronic vehicle controls "across the entire automotive industry." In addition, NHTSA has enlisted National Aeronautics and Space Administration (NASA) engineers who are experts in software and electronics to examine the potential for problems with computer hardware or software or electronic systems in increasingly complex automobiles.[4]

## NHTSA

To ensure that federal regulators receive early notification of potentially dangerous flaws in motor vehicles, and that auto manufacturers quickly take corrective action, Congress in 2000 passed the Transportation Recall Enhancement, Accountability and Documentation (TREAD) Act (P.L. 106-414).[5] The TREAD Act, developed in response to deaths stemming from flawed Firestone tires, created an "Early Warning" system requiring automobile vehicle and equipment manufacturers to promptly report data to NHTSA regarding defects, injuries, or deaths from auto vehicles or equipment. The law also required companies to report to NHTSA any recalls of its products in other countries within five days and imposed tougher penalties for manufacturers that did not comply with the act. The TREAD law was also designed to give consumers more timely and complete information about possible automotive safety issues.[6]

One of Congress's intentions in passing the TREAD Act was to provide NHTSA with tools to identify potential defects more quickly. Some consumer groups and industry experts say NHTSA is still underfunded and does not have enough power to quickly force recalls, forcing it to instead negotiate with auto companies on less-sweeping sanctions. Critics, such as former NHTSA Administrator Joan Claybrook, also say that Toyota and NHTSA have not responded aggressively to reports of sudden acceleration. NHTSA officials reply that they acted promptly and repeatedly in response to information at hand. But, nonetheless, NHTSA in April 2010 levied a $1 6.375 million fine against Toyota—the largest allowed under law—charging that the company did not comply with the TREAD Act requirement to promptly report safety problems to the agency.[7]

The most pressing issue, however, is that NHTSA[8] inspectors have been unable to identify a cause or causes that would explain all the incidents of sudden or unintended acceleration. Toyota officials also say they have not found a mechanical or electrical problem that would account for complaints that do not seem to arise from defective floor mats or sticky accelerator pedals, which have been the subject of recalls. The net result is that there is currently no solution, or even explanation, for many of the reported incidents.

The DOT and Toyota officials are working to address the sudden acceleration complaints. Toyota has announced the creation of a special panel, headed by former Secretary of Transportation Rodney Slater, to examine quality and safety issues. Toyota will appoint, as part of this response, a safety officer in every country in which it has operations, and will open technical offices in some U.S. cities. The company has also said it will make brake override systems (software that gives priority to braking when the brake and accelerator are depressed at the same time) standard on new model cars.[9] The company is installing brake override systems in recalled models, when possible.

Consumer groups are pressing Toyota and NHTSA to look more closely at the possibility that the incidents may have been caused by defects in the vehicles' electronic systems.[10] Both NHTSA and Toyota officials say their research has, so far, not detected software or electronic issues, particularly with electronic throttle controls. In addition to LaHood's announcement that the DOT has asked for studies from the NAS and help from NASA engineers, the Transportation Secretary has asked the DOT Inspector General (DOTIG) to determine whether NHTSA conducted an adequate review of alleged unintended acceleration and has the necessary staff. He has also indicated that the Obama Administration is considering requiring all auto manufacturers that produce cars with electronic throttle controls to install an emergency brake override on their vehicles. Such systems are already standard on certain makes of automobiles.

Congress is now considering amending the TREAD Act or taking other steps to enhance NHTSA's power and, as a result, public safety. Senate Committee on Commerce, Science, and Transportation Chairman Jay Rockefeller of West Virginia said during a March 2, 2010, hearing that he wants to proceed expeditiously with legislation. In the House, the Committee on Energy and Commerce and the Committee on Oversight and Government Reform have launched their own investigations into the sudden acceleration cases, and the response of Toyota and NHTSA. Lawmakers are also considering the FY2011 budget for DOT, including NHTSA. Secretary of Transportation Ray LaHood told Congress that President Obama's FY2011 budget includes authority for an additional 66 full-time equivalent positions at

NHTSA.[11] There are currently 57 full-time NHTSA employees devoted to safety investigations. DOT officials have said they will determine how many of the new positions will be used for safety investigations.

## FEDERAL OVERSIGHT OF AUTO SAFETY

NHTSA is the federal agency with primary responsibility for regulating motor vehicle safety, including setting motor vehicle safety standards. It is also the main agency charged with identifying motor vehicle safety defects through its Office of Defects Investigation (ODI). The ODI receives complaints from consumers and certain information from manufacturers, assesses the information to determine whether there appear to be trends indicating potential safety defects, and conducts investigations to determine whether the evidence indicates the presence of problems. When a safety defect is identified, manufacturers are required to provide owners of the affected vehicles with a remedy at no cost to the owners.

Congress, over the years, has attempted to strengthen NHTSA's authority, most recently with the October 2000 passage of the TREAD Act. The measure was drafted out of concern that NHTSA did not respond quickly enough to a number of crashes and deaths that ultimately led to the recall of millions of Firestone tires.[12] The TREAD Act included provisions to increase the amount of data available to NHTSA regarding potential safety defects, and authorized more funding to improve NHTSA's capacity to more quickly analyze data. (See **Appendix A** for additional context and more details.)

The TREAD Act obligates manufacturers to notify NHTSA within five days of determining that a safety defect exists. A key provision is an Early Warning Reporting (EWR) requirement (section 3(b)). The EWR requires manufacturers to regularly share with NHTSA certain information they receive about their vehicles or equipment. Specifically, manufacturers are required to report on a quarterly basis:

1. The number of vehicles, tires, and child restraint systems, by make, model, and model (production) year, that were produced;
2. The number of claims and notices involving death (including those in foreign countries), personal injury, and property damage by make, model, model year, and vehicle identification number;
3. The number of paid warranty claims in the United States that involve specified components and systems;

4. The number of field reports received from the manufacturer's employees, representatives, dealers, and fleets, related to problems with specified components and systems, and copies of all field reports (except those from dealers); and
5. All consumer complaints received regarding a product.

Other provisions require manufacturers to inform NHTSA of safety recalls or other safety campaigns conducted in foreign countries on motor vehicles or equipment similar to those sold in the United States.

The average number of annual automotive defect recalls in the United States has increased from 486 during 2000-2004 to 524 during 2005-2009. (See **Table 3**.) About 22% of all recalls from 2005-2009 were ordered by NHTSA; the rest were voluntary actions announced by auto companies—including the most recent Toyota recalls.[13] Consumer groups say the percentage of government-ordered recalls is small, indicating that NHTSA is not tough enough, and that auto companies can negotiate less onerous penalties. NHTSA Administrator David Strickland told Congress in March 2010 that voluntary recalls are often preferable, since NHTSA-ordered actions can be subject to lengthy delays. For example, if a manufacturer resists a NHTSA recall order, the agency must go to court to prove that a defect exists that creates an "unreasonable safety risk."[14]

NHTSA estimates that about half the investigations it opens result in a safety recall or other manufacturer action.[15] (For further information on NHTSA's Defects Investigation Process, see **Appendix B**.)

## Sudden Acceleration Complaints

Since 2000, NHTSA has received about 30,000 to 50,000 complaints annually about a wide array of potential problems and defects in automobiles. About 2% to 5% of the overall complaints each year have involved reports of possible unintended acceleration.[16] Since 1980, NHTSA has carried out 141 investigations into consumer throttle-control complaints in a variety of automobiles, involving both electronic and mechanical throttles.[17]

Determining the cause of sudden acceleration is often difficult. A problem may occur rarely, for example, or NHTSA may not be able to reproduce a specific incident reported by a consumer. Further, there are several potential causes for sudden acceleration: driver error; design defects that may induce driver error (such as poor pedal placement); the absence of a shift interlock (a shift system designed to make it more difficult for a person to accidentally shift a car into drive); interference from floor mats; mechanical or electromechanical defects; and electronic defects.[18]

Several studies conducted by NHTSA and vehicle safety agencies in other countries have concluded that driver error—inadvertently pressing the accelerator when intending to press the brake—is generally the cause of unintended acceleration.[19] Critics of this explanation in the Toyota case observe that these studies were done on 1980s-era vehicles, and thus are of uncertain value in examining contemporary vehicles that use new technologies.[20]

## Audi Historical Precedents

One of the most prominent studies linking driver error to unintended acceleration arose from a 1980s federal investigation into reported problems in Audi automobiles. In the mid-1980s, NHTSA received a number of complaints from vehicle owners who alleged that their Audi 5000 automobiles had suddenly accelerated as they shifted their automatic transmission from park to drive or reverse. The New York Attorney General and consumer groups requested an investigation.[21] The television program *60 Minutes* aired a segment that purported to show an out-of-control Audi.[22] Audi in 1986 announced a recall and installed a gear shift lock that required consumers to depress the brake pedal in order to shift into gear. Still, U.S. Audi sales plummeted from 74,000 vehicles in 1985 to a low of 12,000 in 1991, before beginning a slow rebound.[23] The resale value of Audi makes, including those not involved in the recall, declined substantially.[24] In 1989, NHTSA released a comprehensive study it had commissioned from an outside research group that examined the Audi incidents, as well as other makes of automobiles, to look for a potential cause of the sudden acceleration. The study concluded that the reported sudden acceleration was caused by driver error.[25]

NHTSA officials have referred to the Audi case as they have looked into recent reports of sudden acceleration in Toyotas and other vehicles. For example, NHTSA in 2003 denied a petition for an investigation into reports that Toyota's Lexus 300 and 400 model autos were prone to unexpected acceleration. In announcing its decision, NHTSA said the data did not provide a "reasonable possibility" that an investigation would result in an order of a safety-related defect and remedy.[26] To reach that conclusion, NHTSA officials examined 182 consumer complaints for three models of vehicles in five

NHTSA categories related to issues of speed control including throttle linkages and cruise control systems. The review covered the period from 1997 to 2000. Based on the data, NHTSA officials determined that Lincoln-make automobiles had a higher complaint rate, 10.6 per 100,000, than the Lexus, 10.0 per 100,000. Cadillac had 8.7 complaints per 100,000 vehicles.[27] NHTSA did a second review limited to just one speed control category. That turned up similar results with 1.2 complaints per 100,000 for Lexus, 1.8 for Lincoln and .76 for Cadillac. In deciding not to proceed with an investigation, NHTSA also referred extensively to the 1989 Audi study linking unintended acceleration to human error.

## Toyota-Related Complaints since 2000

NHTSA data indicate that the agency has received a range of 1,200 to 2,000 complaints annually since 2000 specifically regarding sudden or unintended acceleration in a range of vehicles. Toyota vehicles accounted for about 4% of all such reports in 2000, a share that rose to 17% in 2008 and 33% in 2009. The increased incidence of complaints occurred as domestic sales of Toyota vehicles climbed substantially, meaning Toyota cars made up a larger proportion of all U.S. autos.[28] Of the Toyota complaints involving sudden unintended acceleration during the period 2000 to 2010, 43 involved fatal crashes; these 43 crashes resulted in 52 deaths. Thirty-three of the 43 crashes involved vehicles equipped with electronic throttle controls; fully three-quarters of the reports were received since November 2009, following increased media attention to the sudden acceleration issue (some of the reports apparently referred to crashes that occurred months or years before).[29]

Another way to break down the data is to look at the number of complaints per 100,000 automobiles. According to an analysis by auto industry publisher Edmunds.com, based on 2005- 2009 data, Toyota had 4.81 complaints per 100,000 vehicles sold, compared to Ford with 3.12 complaints. (See **Table 1**.) Toyota has had more complaints of unintended acceleration on its newer-model vehicles—autos sold from marketing year 2005 through September 30, 2009—than any of the other so-called Big 6 auto manufacturers, according to Edmunds.com.[30]

State Farm Insurance Co. reported to NHTSA in 2004 that it had seen an increase in insurance claims related to unintended acceleration in Camry and Corolla vehicles. State Farm noted that the claims surged beginning with the model year 2002 Camrys, in which Toyota had introduced an electronic throttle. State Farm claims relating to Toyota Corollas also increased after the introduction of the electronic throttle in those vehicles in model year 2005. After an initial spike, claims decreased, though Corolla complaints in 2009 were still generally higher than in the years preceding the change.[31]

**Table 1. Complaints of Unintended Acceleration in Selected Brands Data for Cars Sold in United States from 2005 Through September 2009**

| Manufacturer | Complaints | Sales | Complaints Per 100,000 Vehicles Sold |
|---|---|---|---|
| Chrysler | 156 | 9,082,757 | 1.72 |
| Ford | 339 | 10,864,479 | 3.12 |
| GM | 134 | 16,542,608 | 0.81 |
| Honda | 89 | 7,059,784 | 1.26 |
| Nissan | 50 | 4,657,974 | 1.07 |
| Toyota | 532 | 11,049,513 | 4.81 |

Source: Edmunds.com (httpi/www.edmunds.com/).

Notes: Data is for sales of vehicles MY2005 and Newer. Complaints are for MY2005 and newer received by NHTSA through September 30, 2009.

*Consumer Reports* in 2009 performed its own breakdown of NHTSA data, finding that more than 40% of sudden acceleration complaints filed with the federal agency on model year 2008 vehicles involved Toyota or Lexus vehicles. The magazine said that Ford Motor Co. also had a significant number of complaints relating to acceleration issues. *Consumer Reports'* Auto Test Center and Statistics Department examined 5,916 reports dealing with 2008 models and found 166 cases of unintended acceleration. The complaints ranged across 22 brands, however, including 47 complaints about Toyota and five about Lexus vehicles, and 36 complaints about the Ford F-150 pickup.[32]

## NHTSA Response to Toyota Complaints

Since 2002, NHTSA has conducted eight examinations, including three investigations, into complaints of unintended acceleration in Toyotas. It closed most of these examinations after it was not able to identify a safety defect that could have caused the complaint. (See Toyota recall timeline, **Appendix C**.) An investigation led to a Toyota recall in 2007 to address a body trim issue in Toyota Siennas that could interfere with the accelerator pedal. More recently, two NHTSA examinations led to Toyota recalls in 2009 and 2010 covering floor mats and potentially sticky accelerator pedals. NHTSA's acting administrator traveled to Japan in December 2009 to meet with Toyota officials and impress on them the need to quickly address reported problems. According to press reports, some owners of Toyota vehicles have complained that their vehicles exhibited sudden unintended acceleration after being recalled and repaired.[33]

The DOTIG is also conducting an audit of NHTSA's Office of Defects Investigation. The audit will examine both ODI's process for identifying and investigating potential defects, as well as the office's actions in regard to the Toyota recalls.[34] The last audit of the ODI was published in 2004, examining NHTSA's implementation of the TREAD Act.

The DOT on April 5, 2010, announced that it had determined that Toyota did not comply with legal requirements to notify NHTSA within five days of discovering a safety flaw in its products. Transportation Secretary LaHood announced that officials would seek a $16.375 million penalty against Toyota, the maximum possible, for what it said was a failure to notify NHTSA, for at least four months, of sticky accelerator pedals in some Toyota models.[35] NHTSA said internal Toyota documents showed that the company knew of the problems with the pedals since at least September 29, 2009, when it issued repair procedures to distributors in 31 European countries and Canada. About 2.3 million Toyota vehicles were recalled in the United States in late February for sticky pedal issues. Toyota on April 19, 2010, agreed to pay the penalty. Toyota said in a statement that the company acted to avoid prolonged litigation and to move forward on quality improvements.[36] Toyota on April 19, 2010, also announced it would recall 2010 model Lexus GX 460 vehicles to update their Vehicle Stability Control system.

## TOYOTA OVERVIEW

Japan-based Toyota Motor Co. has grown into an international powerhouse since its founding in 1937, and now manufactures cars and trucks, motors, air conditioning units, factory equipment, and other machinery.[37] Toyota has undergone a particularly rapid period of expansion during the past two decades. Toyota's share of the U.S. auto market nearly doubled between 1998 and 2009.[38] (See **Figure 1**.) In 2008, Toyota moved ahead of General Motors as the world's largest auto company, in terms of annual sales.[39] Toyota has long been known for its "lean" manufacturing process, and the company's ability to produce cars more efficiently than its competitors has given it a competitive edge. Toyota has managed to maintain its productivity during the past several decades even as automobiles have become more complex to design and produce, though the gap between Toyota and U.S. auto firms appears to have narrowed.[40]

Toyota has a major U.S. presence, measured in both sales and production. Toyota began selling its vehicles in the United States in 1957, with the founding of Toyota Motor Sales.[41] It also sells its Lexus brand. In 1984 the firm started domestic U.S. production as part of a joint venture with General Motors. Toyota now has 13 manufacturing locations in North America (including Mexico and Canada) that produce vehicles including the Sequoia, Tacoma, Venza, and Avalon.[42] Toyota employs about 29,000 U.S. workers, and its direct U.S. investment has grown to more than $18 billion. The firm spent more than $22 billion on parts, materials, and components from U.S. suppliers in 2009. Toyota sold 1.77 million cars and light trucks in the United States in 2009. Of that total, about 1.1 million were made in this country and 663,844 were imported.[43]

North America is the largest market for Toyota vehicles. In February 2010, Toyota forecast it would sell 2.05 million cars in North America during its 2010 fiscal year (April-March) out of total forecast global sales of 7.2 million.[44]

Toyota Motor Engineering and Manufacturing, headquartered in Erlanger, KY, carries out design and development for U.S. manufacturing operations. Toyota Financial Services provides loans for car buyers, so-called floor-plan loans that provide dealers needed financing to stock inventory, and handles lease coverage and vehicle service/warranty agreements and other financial services for Toyota, Lexus, and Scion dealers, Toyota forklift and Hino dealers, and Toyota Material Handling dealers and affiliates. Toyota Financial Services employs 3,000 U.S. workers with managed assets in excess of $65 billion.[45]

## Toyota Corporate Structure

As Toyota has expanded, the company has increased its range of products and built manufacturing plants in a number of countries, including the United States. There have been suggestions that Toyota put too much emphasis on expansion at the expense of quality.

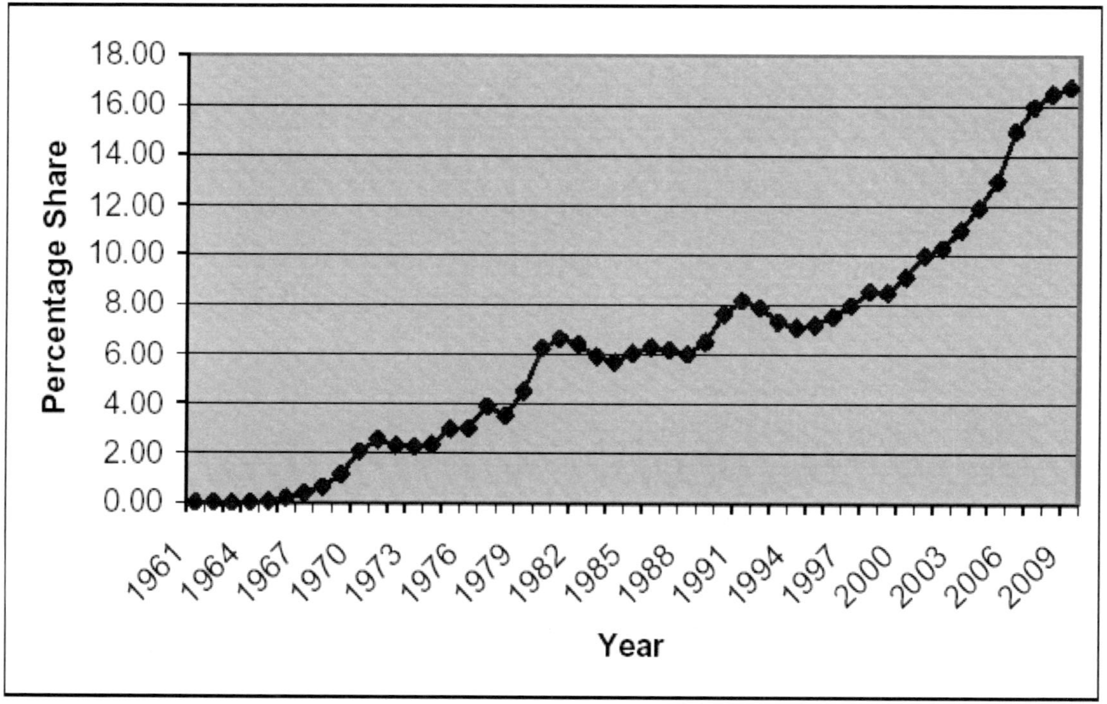


Source: Ward's Automotive Group, *U.S. Vehicle Sales Market Share by Company*, 1961-2009.

Figure 1. Toyota U.S. Auto Market Growth Percentage share of U.S. vehicle sales from 1961-2009.

Further, there have been criticisms that management, which remained centralized in Japan, was slow to respond to problems in its U.S. operations. In a September 2006 internal Toyota presentation entitled "A New Era for Toyota and TMA in North America," then-Toyota Motors North America President Jim Press noted increasing recall activity and federal investigations. Press said that quality problems affected customer loyalty and suggested that the company's credibility with NHTSA was slipping. Notes accompanying the slides in Press's Power Point presentation included this entreaty to corporate managers: "We ask TMC to trust our judgment when we need your urgent help in getting issues resolved. We need faster information flow, and more technical support when hot issues arise."[46]

James Lentz, president and COO of Toyota Motor Sales, USA, in February 23, 2010, testimony before the House Energy and Commerce Committee said that Japanese officials had been making the key decisions on recalls, but said the process was changing.[47] Toyota CEO Akio Toyoda, a grandson of Toyota's founder who became CEO in 2009, has personally apologized to customers, saying the corporation may have grown too big, too fast and, in the process, lost its focus on quality and customer response. Toyoda has proposed changes in Toyota's corporate structure to ensure it responds more quickly and openly to safety and quality questions in the future.[48]

Toyota officials have laid out a series of steps that they say should improve product quality and consumer satisfaction:

- Establish a Special Committee for Global Quality, to be led by the president of Toyota. The panel will review operations and suggest needed changes.
- Create an outside panel of industry experts to evaluate any new quality controls within Toyota. The panel will be headed by former U.S. Secretary of Transportation Rodney Slater.
- Create a global system to better share information within the company and with regulators, including NHTSA.
- Set up teams of technicians in select U.S. cities to quickly respond to reports of unintended acceleration in Toyota vehicles and make on-site inspections.
- Create a new U.S. position of regional product safety executive. Corporate officials have also said they will give North American Toyota managers more autonomy to make decisions about safety and defect issues.

- Establish an Automotive Center of Quality Excellence in the United States.[49]

## Black Box Availability

Toyota is also pledging to become more open in another area. The company has been faulted for making it difficult for U.S. investigators and drivers to gain access to electronic "black box" data for its vehicles that might help to determine the causes of some of the incidents.

In many automobiles, an event data recorder (EDR), commonly referred to as a "black box," records information related to a vehicle crash. An automobile "black box" is different from a commercial aircraft's flight recorder (also popularly referred to as a "black box"), which records both equipment data and audio transmissions. An automobile "black box" continually records information from a vehicle's brakes, seat belts, and air bags; the recording is of short duration, and the recorded information is continuously overwritten by new information, so that at any time there is usually only a few seconds of data recorded. The EDR will store information about the condition of the brakes, seat belts, and air bags for a few seconds before and during a crash. Some EDRs also record data such as engine throttle status and vehicle speed.[50] EDRs began to appear in cars as early as 1991; most cars now have an EDR. NHTSA has been collecting EDR data since 1991 because of its value in helping investigators reconstruct vehicle crashes. In 2006, NHTSA issued a regulation to standardize the data collected by EDRs. The regulation takes effect in 2012, affecting model year 2013 cars.[51] While the rule will require that EDRs record 15 types of data, including throttle status and vehicle speed, it applies only to cars that have EDRs; the regulation does not require that cars be equipped with EDRs.

Toyota engineers describe their EDRs as essentially prototype parts that were designed as event recorders. "There have been times—there have been occurrences where we have been able to download some information regarding throttle angles, steering angle, brake application, et cetera, that does give us some information. But that's very, very limited, and hasn't really occurred that often," John Hanson, national environmental safety and quality communications manager for Toyota said in a recent presentation.[52]

Toyota until recently had just one device in the United States capable of reading its vehicles' proprietary black box data. By contrast, other automobile companies make their black box data more accessible to consumers and investigators. In hearings before Congress, Toyota executives pledged that they would provide 100 such devices. Toyota is also stepping up efforts to improve black box technology on its upcoming models of automobiles.[53]

## Recall Impact on Toyota Sales

Even before the recent recalls Toyota, like other automobile companies, had been hurt by the economic downturn and credit crunch. Toyota officials reacted with aggressive marketing in the wake of the bad publicity surrounding the sudden acceleration incidents. On March 2, 2010, Toyota announced a series of financial incentives through April 5, 2010, designed to lure buyers into its dealer showrooms. The company offered 0% financing or lower monthly lease rates on some popular models, as well as a complimentary two-year maintenance program for customers who buy or lease a new car and provide proof that they already own a Toyota make of vehicle.[54]

Toyota Motor Sales, USA reported on April 1, 2010, that it sold 186,863 automobiles in March, a 35.3% increase compared to the same period in 2009.[55] For the first quarter of 2010, sales were up 8.7% from year-ago levels, measured in sales per day. Sales rose 7.2% on an unadjusted volume basis—the difference comes from the fact that there were 74 selling days in the first quarter of 2010 compared to 75 in the first quarter of 2009. Still, some analysts have estimated that Toyota could face billions of dollars in losses from the sudden acceleration cases.[56]

## Toyota Lobbying

In addition to technical issues, lawmakers have questioned whether Toyota lobbyists were able to persuade NHTSA officials not to order recalls in certain instances.

In 2009, Toyota had a larger lobbying force in Washington, DC, than any other foreign automobile company. According to the nonprofit Center for Responsive Politics, Toyota and its subsidiary organizations spent more than $5 million and employed 31 lobbyists in 2009.[57] By comparison, Nissan Motors, another large foreign auto firm, spent $3.67 million and employed 19 lobbyists, while Honda had nine lobbyists and spent $2.56 million during the period.[58] Honda, for example, also has a major U.S. presence. The company opened its first U.S. plant in 1979 and now directly employs more than 25,000 Americans.[59]

Lawmakers have focused on a July 6, 2009, document prepared by Toyota's Washington office for Toyota Motors COO Yoshi Inaba. In that document, the office claimed a number of "wins" including delaying final safety rules by NHTSA, and persuading NHTSA officials to impose lesser defect sanctions, including a negotiated equipment recall on Camrys that saved the company $100 million. Inaba told lawmakers during congressional hearings that the document does not reflect official Toyota policy and that he does not know where the $100 million figure came from.

A number of former DOT employees now work for Toyota and other auto companies.[60] Secretary of Transportation Ray LaHood told Congress that a review of internal records indicates that the former officials have complied with laws that limit their ability to lobby their former colleagues. The Transportation Secretary has suggested tightening the lobbying laws, however.

# Policy Issues and the Congressional Response

The increase in consumer complaints regarding sudden acceleration in Toyota vehicles, and the government and industry response, has led some Members of Congress, and consumer and industry groups to examine a number of issues regarding the specific cases at hand—and the broader issue of government regulation of automotive safety.

## Does NHTSA Have Enough Resources for Defects Investigation?

Some Members of Congress and consumer groups have suggested that NHTSA lacks the resources to effectively carry out its responsibility to detect automotive defects. Joan Claybrook, a former head of NHTSA, says the agency's budget and staffing were reduced by almost one- third during the 1980s, and have never been restored to earlier levels.[61] However, while that may be true of NHTSA as a whole, it does not necessarily apply to the defects investigation work of NHTSA. NHTSA's budget for defects investigation, adjusted for inflation, was larger in FY20 10 (at $9.8 million) than in FY1981 ($4.7 million).[62] The number of staff dedicated to defects investigation has increased to 57 in FY2010 from 45 in FY2000, following passage of the TREAD Act.[63]

Whether NHTSA's budget for defects investigation is sufficient for protecting the public from unreasonable safety defects is another question. No specific evidence has been cited in support or in contradiction of the claim that NHTSA lacks sufficient staff resources to conduct needed investigations. Some have pointed to NHTSA's repeated reference, in declining to open defect investigations after examining defect petitions, or in terminating defect investigations without having identified a safety defect, to its "need to make effective use of limited resources," suggesting that language indicates that NHTSA does not have sufficient resources.[64] But the reference to making effective use of limited resources is an acknowledgement that, at some point during an investigation, if no defect has been identified, the investigation needs to be concluded so that investigators can turn to other investigations.[65]

The DOT Inspector General, in a 2002 review, questioned the limited amount of time staff had to assess consumer complaints. In its review, the DOTIG estimated that the staff responsible for reviewing the roughly more than 34,000 annual complaints had "an average of about 12 minutes per complaint to review the information, search the defect database for similar complaints, related investigations, and recalls, and decide whether to recommend an investigation."[66] The DOTIG report also noted that the complaint database was filled with errors. This problem appears to persist: one group that reviewed NHTSA's complaints found the complaint database "a mess. After reading each complaint since model-year 2005, we found that 30 percent of the original complaints were miscategorized; more than 26 percent were duplicates."[67] NHTSA reports that currently all complaints are initially screened and entered into NHTSA's complaints database by a pair of screeners. These two screeners' responsibilities include categorizing the complaints for further review by a group of six screeners with various areas of expertise who review complaints that fall into their specialties.

DOT officials have testified that the Administration's FY2011 budget requests 66 additional positions for NHTSA.[68] They have not indicated how many of these new positions might be dedicated to defects investigations, saying that the allocation of the positions within NHTSA will be based on need. NHTSA's FY2011 Budget Estimate shows that the requested FY2011 funding for Safety Defects Investigations is the same amount enacted for FY2010: $9.829 million.[69] The total FY2011 budget request for NHTSA was $878 million, $5 million more than the FY2010 enacted level. Most of NHTSA's funding (71%—$620 million of the $873 million total for FY2010) is distributed to states in the form of grants to support traffic safety activities. Driver behavior programs are implemented by the states, not the federal government, which is why most of NHTSA's funding goes to states' efforts to change driver behavior.

NHTSA's mandate is the promotion of auto safety, usually measured by the reduction of traffic deaths and injuries. The vast majority of traffic deaths and injuries result not from vehicle safety defects, but from driver behavior: driving while intoxicated, driving while distracted, speeding, vehicle occupants not wearing seat belts, motorcyclists not wearing helmets.[70] **Table 2** indicates the relative scale of crashes and fatalities possibly attributed to reports of sudden acceleration in Toyotas in the context of traffic deaths from all causes. Some people contend that NHTSA's defects investigation resources should be increased in order to promote traffic safety; other observers contend that any additional resources would have a greater impact on safety if applied toward driver behavior programs.

**Table 2. Scale of Toyota Sudden Unintended Acceleration Incidents Data for Toyota, 2000-2010; Totals 2000-2008**

| | Attributed to Sudden Unintended Acceleration in Toyota Vehicles | Total from All Causes |
|---|---|---|
| Crashes involving Fatalities | 43 | 340,152 |
| Resulting Fatalities | 52 | 377,604 |
| Resulting Injuries | 38 | NA |

Source: Toyota data: NHTSA, cited in "Toyota Sudden Acceleration Tied to 43 Fatal Crashes," by Angela Greiling Keane and Jeff Plungis, Bloomberg.com, March 2, 2010; Total crashes and fatalities: NHTSA, *Traffic Safety Facts 2008*, Tables 1 and 2.

Notes: The *Traffic Safety Facts* report has figures for both deaths and injuries that result from crashes, but does not report a number for the subcategory of injuries that resulted only from those crashes that also involved fatalities.

Another point raised in connection with the Toyota issue is that NHTSA may not have the right mix of investigative expertise. Specifically, in view of the increasing role of electronics and computer software in automobiles, the question has been raised as to whether NHTSA has sufficient numbers of electrical and software engineers for its investigations. NHTSA officials have testified that they have five electrical engineers on staff, plus another electrical engineer and a software engineer at the Vehicle Research and Test Center in

East Liberty, OH.[71] Subsequent information from NHTSA indicates that two of the electrical engineers work in the Office of Defects Investigation, whose staff includes a total of 26 engineers.[72]

**Table 3. Motor Vehicles and Equipment Defect Recall Trends Total Defect Recalls and NHTSA-Influenced Defect Recalls**

| Period | Total Defect Recalls | | NHTSA-Influenced Defect Recalls | | | |
|---|---|---|---|---|---|---|
| | Average Annual # of Recalls | Average Annual # of Items Affected | Average Annual # of Recalls | Average Annual # of Items Affected | % of Recalls | % of Items Affected |
| 1995-1999 | 306.8 | 23,605,522 | 80.4 | 12,249,817 | 26.2% | 51.9% |
| 2000-2004 | 485.8 | 25,739,544 | 95.6 | 15,765,715 | 19.7% | 61.3% |
| 2005-2009 | 524.4 | 17,078,418 | 118.6 | 8,627,948 | 22.6% | 50.5% |

Source: Data from NHTSA, Recall Management Division, updated as of January 11, 2010.

Notes: "Items Affected" includes vehicles, auto equipment (e.g., tires), and child safety seats.

## Is the Toyota Issue a Sign of Broader Problems within the Auto Industry?

In general, contemporary cars are safer and more reliable than ever before.[73] But as a result of these improvements, cars are also more complex than ever. As **Table 3** indicates, the number of automotive recalls per year has generally been increasing, although the number of vehicles, tires, and equipment affected by these recalls has been declining. Most recalls are initiated by manufacturers, not in response to NHTSA action.

The trend of increasing numbers of recalls is seen in other countries as well. For example, a study of recall actions in the United Kingdom found that the number of recalls per year rose between 1992 and 2002.[74] The reason(s) for the increasing number of recalls is not clear. Among the explanations that have been proposed are, for example, that (1) the increasing complexity of automobiles, combined with a quicker product development cycle in the automotive industry, are resulting in cars coming to market with more safety-related problems than in the past; and (2) that safety-related defects are no more frequent than in the past, but concerns about lawsuits and greater scrutiny have made manufacturers more inclined to recall vehicles for issues that earlier might have been ignored.

## Are Electronics and Software Testing Stringent Enough?

An emerging issue for lawmakers and investigators is the possibility of problems arising from (1) glitches with software used to run onboard systems, (2) problems in computer hardware, or (3) issues regarding internal electronic compatibility and outside electromagnetic interference that can impede the performance of automobiles.

### *Electronics*

Automobile manufacturers perform multiple tests of their products to ensure that the electronic systems in their vehicles work as intended. In general, automotive companies have developed standards, equipment, and techniques to test whether electronic systems are vulnerable to a range of possible forces that could impair their operation:

- The inner workings of the automobile—each vehicle has internal electronic components that have their own circuitry and could potentially create interference. The auto ignition system is one major energy source.[75]
- Static interference—as an automobile moves down the road, for example, it experiences high-speed friction with the surrounding air that can produce static electricity.
- Outside interference—a traveling car or truck passes a variety of outside electromagnetic forces such as high-voltage transmission lines or television and radio communication equipment.[76]

According to Todd Huber, professor of vehicle electronics at the Clemson University International Center for Automotive Research, most automobile manufacturers carry out component- and vehicle-level testing of electronic systems. Most of the relevant tests are based on international standards, many of which began as standards created by the SAE International, an association of engineers and technical experts in aerospace, automotive, and commercial-vehicle industries. The electronics standards are usually voluntary and are not legal requirements. Manufacturers often have their own standards, which

Huber says may differ from the international standards.[77] Clemson has compiled a list of relevant automotive electronics standards.[78]

Toyota has eight special test chambers for electronics in Japan. The company began building a testing chamber in Michigan two years ago, though it will not be completed for another two years. Toyota engineers say they test their components based on European Union, Society of Automotive Engineers, or international standards. For example, for testing potential interference from low-frequency electronic devices, the company uses EU standards.[79] The company also has its own standards.

### *Software*

Software applications were first introduced into car ignition systems 30 years ago. Today, electronics and software can account for 40% or more of the production cost of a car, and a high- end automobile can require more than 10 million lines of computer code to run hundreds or thousands of individual functions and dozens of electronic components.[80] Some cars, with global positioning systems, wireless communications, cruise control, automatic parking control, and a slew of engine-related functions, have software code that begins to rival that of more sophisticated equipment, like aircraft.

Automobile manufacturers procure parts from outside suppliers, and then create finished vehicles assembly-line style. Some experts say that such a system may not be optimal for vehicles that now contain dozens of data processors that control thousands of software functions and where phenomena like unintentional interaction between different software and hardware components can become issues.[81] Long-term reliability data may not be available for some software.[82] Owners may not service their cars as frequently and rigorously as recommended, though software is designed to alert drivers to needed tuneups or unexpected problems with the automobile.[83] Auto companies try to design their products to account for tough driving conditions.

Auto manufacturers have experienced glitches with software. For example, Toyota in February 2010 announced it was recalling certain models of its Prius-make vehicles to update software controlling the anti-lock braking system.[84] Other auto companies have recalled millions of cars for software flaws ranging from problems with air bag deployment to the failure of software controlling onboard music and other entertainment systems.[85] Some experts point out that while there have been high-profile cases of automobiles recalled for software problems, mechanical flaws remain a much larger problem for the industry.[86]

Software development is becoming an increasingly expensive area for auto manufacturers, which is leading to expanded efforts to increase standardization. Another factor influencing industry moves to systematize elements of software design is heightened safety concern as onboard computers begin to carry out functions that more directly affect the control of an automobile.[87] Vehicles have advanced from single-control devices like airbags, to braking systems where several components are linked into a network to systems that use several computer networks and also are also linked to systems outside the car through a radio, for example. [88]

Automobile manufacturers use a variety of standards in developing software, including guidelines developed by the Motor Industry Software Reliability Association (MISRA), an international automotive industry association based in the United Kingdom.[89] The MISRA guidelines are designed to ensure that software code is standardized and dependable and also offers guidance on safety analysis.[90] The MISRA guidelines are widely used in other transportation systems, including aviation and railroads.

The Society of Automotive Engineers recently announced enhanced software guidelines, including the creation of a repository for information about the industry, best practices for software design, and criteria for evaluating industry compliance with best practices.[91] Manufacturers will be able to access the repository, when it is up and running, for information on the credentials of designers working in various areas. The repository will also allow companies to share data.

Another industry effort is Autosar, or Automotive Open System Architecture, which was formed in 2003. Autosar is a global coalition of automobile manufacturers, suppliers and electronics, semiconductor, and software firms working to develop standardized software for the auto industry. The group's "core partners" include BMW Group, Bosch, Daimler, Ford, General Motors, Toyota, and Volkswagen. Autosar describes its mission as "simplifying the exchange and update options for software and hardware" to control the complexity of electrical and electronic systems in motor vehicles.[92] The industry group says increased standardization should reduce costs over the long run, allow companies to focus more on specific high-end applications and make it easier for suppliers to create products that can be used across the industry.

### *Electronic Throttle Problems*

While Toyota and NTHSA have identified two mechanical causes for the sudden acceleration— improperly designed floor mats and "sticky" accelerator pedals—some automotive experts and lawmakers say officials have been too quick to dismiss the possibility that the cars' electronics systems could be a factor.

Electronic throttle control, often described as drive-by-wire, is a system where sensors, rather than a mechanical cable or rods attached to an accelerator, are used to control the throttle. In such systems, the sensors send signals to onboard computers that determine how far to open the throttle, which determines air flow and fuel usage. The system, by improving efficiency, can increase mileage and reduce emissions and provide faster response time and better maneuverability.[93]

Electronic throttles are engineered with redundant components in an effort to create a reliable, fail-safe system and prevent unintended acceleration or other potential problems. Toyota says its system is designed with multiple layers of technology that should protect a driver in case of unanticipated difficulties. For example, Toyota accelerators and throttles each have two sensors that must work together for the car to operate. In the case of a serious malfunction, the vehicle is engineered to revert to idle mode.[94] Automakers install indicator lights that illuminate in the event of a potential malfunction, alerting a driver of a potential problem. In many cases, according to automakers, the onboard computer controlling the throttle also receives a diagnostic trouble code when a problem is detected. The code is stored in the computer memory, aiding in repair of the vehicles or in reconstruction of events leading up to an incident.

Some automobiles include a brake override system that essentially halts power to the car if the accelerator and brake pedal are both depressed at the same time. Toyota is installing the brake override software in autos brought back on recall, if the particular vehicle is equipped to receive the update. Toyota will start installing the equipment as a standard feature on 2011 model year vehicles.[95]

While NHTSA and Toyota have not to date discovered a problem with the electronics in the cars they have tested for sudden acceleration, they are expanding their investigations. Specifically, the DOT has asked the National Research Council (NRC) to spend 15 months looking at both instances of unintended acceleration and to examine electronic vehicle controls across the entire automotive industry. The effort will include a review of previous industry and government efforts to identify possible sources of unintended acceleration as well as automotive software and computer hardware design, electromagnetic compatibility, and potential electromagnetic interference. The panel is to make recommendations on how to improve auto safety in these areas.[96]

NHTSA's separate review of electronic throttle controls in Toyotas, aided by NASA engineers, is expected to be completed by late summer 2010. The DOT has announced that nine NASA experts will aid NHTSA, and more could join the effort if necessary. Some outside experts have already performed their own analysis of Toyota engineering. Toyota is pushing back, for example, against a study by Southern Illinois Professor David Gilbert in which he said he had discovered a flaw in the electronics of a Toyota vehicle that could lead to unintended acceleration.[97] Toyota on March 8, 2010, charged that Gilbert had rewired and re-engineered the car in ways that were "virtually impossible to occur in real-world conditions." Toyota officials announced that they had been able to produce similar results on automobiles manufactured by other companies, using Gilbert's techniques.[98]

# Congressional Options

Congress in February 2010 began a series of hearings on the issue of unintended acceleration in Toyota vehicles. Lawmakers have requested thousands of pages of documents from Toyota and NHTSA, and have indicated that they plan to hold additional hearings. As part of the congressional examination, lawmakers are examining the actions of Toyota and NHTSA in the period leading up to the recent, extensive recalls.

Senate Commerce Committee Chairman Jay Rockefeller of West Virginia has laid out possible steps for Congress and the DOT, including taking a fresh look at the TREAD Act, requiring brake override technology on all new automobiles, requiring auto makers to provide the necessary equipment for investigators to read electronic data recorders and forcing senior executives of auto companies to personally certify that information their firms provide to NHTSA is complete and accurate.[99]

Secretary of Transportation Ray LaHood also has raised the possibility of requiring brake override systems in all new cars, and has suggested that Congress tighten lobbying laws and regulations to make it more difficult for former officials to lobby their former agencies. The DOT has pulled in NASA engineers to take a closer look at potential issues with electronics and is continuing to assess whether Toyota has reported possible product defects in an expeditious manner, as required by law.

Consumer advocates have suggested broader changes including new standards for accelerators and electronics testing; mandatory installation of event data recorders and the collection of more information; more public disclosure of Early Warning Reports under the TREAD Act; higher fines for violations of the TREAD law; and enhanced funding for NHTSA.[100]

# APPENDIX A. NHTSA AND THE TREAD ACT

In October 2000, Congress passed the TREAD Act (P.L. 106-414). The act was passed in response to a safety problem involving certain Firestone tires, which were linked to a number of crashes and deaths, and which led to a recall of millions of tires. During hearings on this issue, Congress examined why NHTSA had not spotted the tire problems sooner, and whether there were changes that could help NHTSA more quickly identify possible safety defects. Congress concluded that NHTSA did not have sufficient information to identify the problem with the tires, but also that NHTSA had not made effective use of the limited information it did have. [101] In the TREAD Act, Congress authorized provisions to increase both the amount of data available to NHTSA to help it to identify potential safety defects, and resources to improve NHTSA's capacity to analyze data to more quickly identify possible defects.

## Increasing Potential Defect Information Available to ODI

Section 3 of the TREAD Act included new reporting requirements for motor vehicle and equipment manufacturers. One provision required manufacturers to inform NHTSA of safety recalls or other safety campaigns conducted in foreign countries on motor vehicles or equipment similar to those sold in the United States. The law also gave manufacturers an obligation to notify NHTSA within five days of determining whether a defect exists. A key provision was the Early Warning Reporting (EWR) requirement (section 3(b)). The EWR requires manufacturers to regularly share with NHTSA certain information they receive about their vehicles or equipment. Specifically, manufacturers are required to report on a quarterly basis

1. The number of vehicles, tires, and child restraint systems, by make, model, and model (production) year, that were produced;

2. The number of claims and notices involving death (including those in foreign countries), personal injury, and property damage by make, model, model year, and vehicle identification number;
3. The number of paid warranty claims in the United States that involve specified components and systems;
4. The number of field reports received from the manufacturer's employees, representatives, dealers, and fleets, related to problems with specified components and systems, and copies of all field reports (except those from dealers); and
5. All consumer complaints received regarding a product.

Prior to the implementation of the EWR requirement, NHTSA's primary information sources for potential vehicle safety defects were technical service bulletins from manufacturers, recalls conducted by manufacturers, and complaints by consumers directly to NHTSA, as well as NHTSA's own defect investigations. While NHTSA was receiving 40,000 to 50,000 consumer complaints each year, that was a small fraction—on the order of 10%—of the number of complaints that were received by manufacturers themselves.[102]

One issue that arose during implementation of the EWR provision was the question of to what extent NHTSA would make public the data provided by manufacturers. An auto equipment industry group sued to require NHTSA to withhold all manufacturer data except when an investigation was initiated, on the grounds that the information could subject companies to competitive harm, and that it was not the intention of Congress that such data should be made public. Conversely, a public safety group sued to require NHTSA to make all of the data available, on the grounds that safety defects could be more effectively identified if the public was also able to view all the information. NHTSA took a middle course. It makes certain data it receives from the manufacturers publicly available (light vehicle production numbers and death, injury, and property damage incidents by make, model, model year, and reported components involved), while keeping other data confidential (e.g., warranty claims, consumer complaints to the manufacturer, field reports).[103] NHTSA's reasoning was that too much disclosure could create a competitive disadvantage for manufacturers (by providing certain information to competitors), and could impair NHTSA's ability to obtain similar information in the future (since manufacturers might limit their information-gathering efforts if they were required to disclose all that information).[104]

## Increasing ODI's Capacity to Analyze Potential Defect Information

After passage of the TREAD Act, NHTSA replaced the ODI's defects information system with a new system, ARTEMIS (Advanced Retrieval (Tire, Equipment, Motor Vehicle) Information System). ARTEMIS was conceived as an information system that could both compile the greatly increased amounts of information NHTSA would receive as a result of the EWR requirement and provide sophisticated data-mining capacities for analysis of this information.

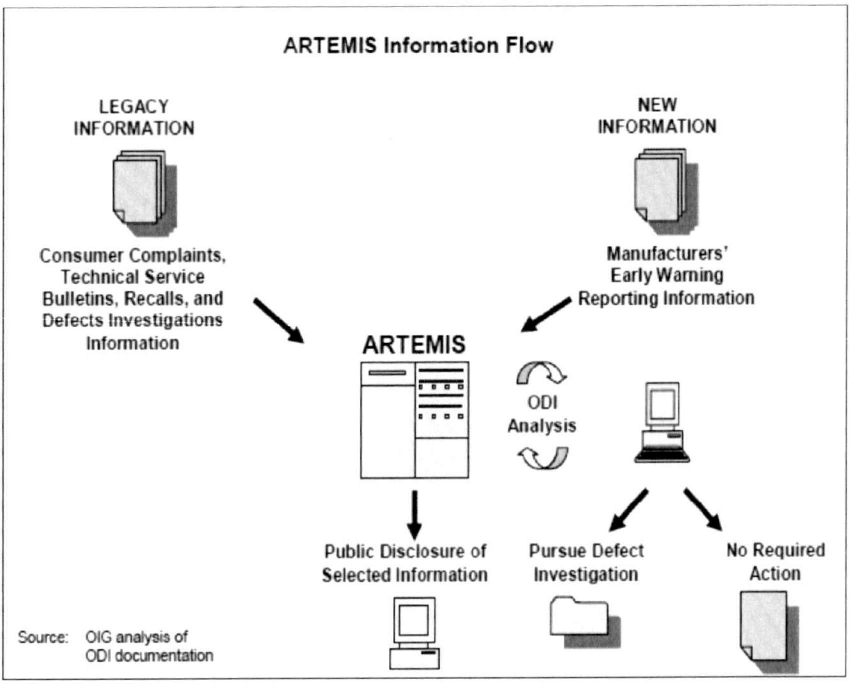


Source: United States Department of Transportation, Office of the Inspector General, *Follow-Up Audit of the Office of Defects Investigation*, MH-2004-088, September 23, 2004, p. 28, http://www.oig.dot.gov/sites/dot/files/ pdfdocs/mh2004088.pdf.

Figure A-1. ARTEMIS Information Inputs.

ARTEMIS's initial 21-month completion schedule stretched to 42 months; its budget nearly doubled (from $5.35 million to $9.4 million); and when it was fully deployed in July 2004, it lacked some of the "more advanced trend and predictive analyses [capabilities] that were originally envisioned as being needed to identify defects warranting investigation."[105] One example of a missing feature in ARTEMIS was the system's lack of ability to automatically highlight trends such as increases in the number of complaints and warranty claims related to a specific motor vehicle component.[106]

In its 2004 audit, the DOT IG was told by NHTSA that it was considering the use of other software to supplement the analytical capacities of ARTEMIS, and also that NHTSA had signed an agreement with the Federal Aviation Administration (FAA) to the effect that the FAA would help NHTSA develop a system to review EWR reports from manufacturers and to identify advanced data analytical options that might be employed.[107] CRS has been unable to obtain updated information about the status of these efforts.

Elements of ARTEMIS are available to the public for searching some EWR information (see http://www-odi.nhtsa.dot.gov/ewr/). According to NHTSA, there are over 9,500 unique visitors to the ARTEMIS site each day.[108] See **Figure A-1** for an illustration of the information inputs for ARTEMIS.

# APPENDIX B. NHTSA'S DEFECTS INVESTIGATION PROCESS

NHTSA's defects investigation process has several stages, with points where a decision to begin or continue an investigation is made (see **Figure B-1** and the summary below). In a 2002 review the DOTIG examined a random sample of 59 defect cases that were either recommended for, or led to, investigation. Of those 59 cases, 21 began immediately as investigations, bypassing the defect analysis/investigation recommendation stage. The remaining 38 cases were recommended for investigation. Investigations were opened for 28 of the cases. At the time those decisions were made, there was no process for recording the reasons why investigations were not undertaken. ODI officials were able to provide reasons why at least some of the cases did not lead to investigations. But the DOTIG found no consistent differences between the cases that led to investigations and those that did not.[109]

The DOTIG noted that ODI defect cases for which defects analysts recommended an investigation be started but for which investigations were not started remain open for monitoring. However, the DOTIG observed that there was no system for tracking whether such cases were in fact continuing to be monitored (e.g., no status or progress reports were required). As a result, ODI managers did not know whether staff continued to search for new complaints related to those cases, or whether new information suggested that a case should be re-evaluated for investigation. The DOTIG review identified subsequent complaints in 7 of 10 cases randomly sampled, and found significant increases in the number of complaints received in two of those cases. ODI had not opened an investigation for any of the 10 cases.[110]

## Receiving Information about Potential Defects

The Office of Defects Investigation (ODI) receives information about possible defects through several channels. It maintains a toll-free hotline (888-327-4236) and an online site (http://www.safercar.gov) through which consumers can report complaints about motor vehicles. Consumers can also mail a letter or fax information to the ODI. NHTSA has a form on its website to help elicit information from complainants. NHTSA's website tells visitors that there are four categories for safety complaints: motor vehicles, tires, non-original vehicle equipment (e.g., lights, jacks, wipers, etc.), and child restraint equipment. NHTSA also receives EWR information from manufacturers. This information is entered into the ARTEMIS information system for analysis.

## Defects Analysis and Identification

Complaints are assessed by NHTSA's Defects and Recall Information Analysis Division. In 2004, the DOTIG reported that the process of analyzing information for evidence of safety defects can take one to three months. This may include field investigations, surveys, and testing. When a pattern suggesting the possible existence of a safety defect is found, an analyst is to assemble an initial evaluation (IE) package containing pertinent information about the suspected defect. The IE is to be reviewed by the chief of the Defects and Recall Information Analysis Division. The IE is then to be sent to members of a peer review panel two weeks before a regularly scheduled review panel meeting to consider the evaluation. According to NHTSA, on average each analyst evaluates and presents 35 cases of potential defects each year; this number has been increasing over time. (See **Table B-1**.)[111]

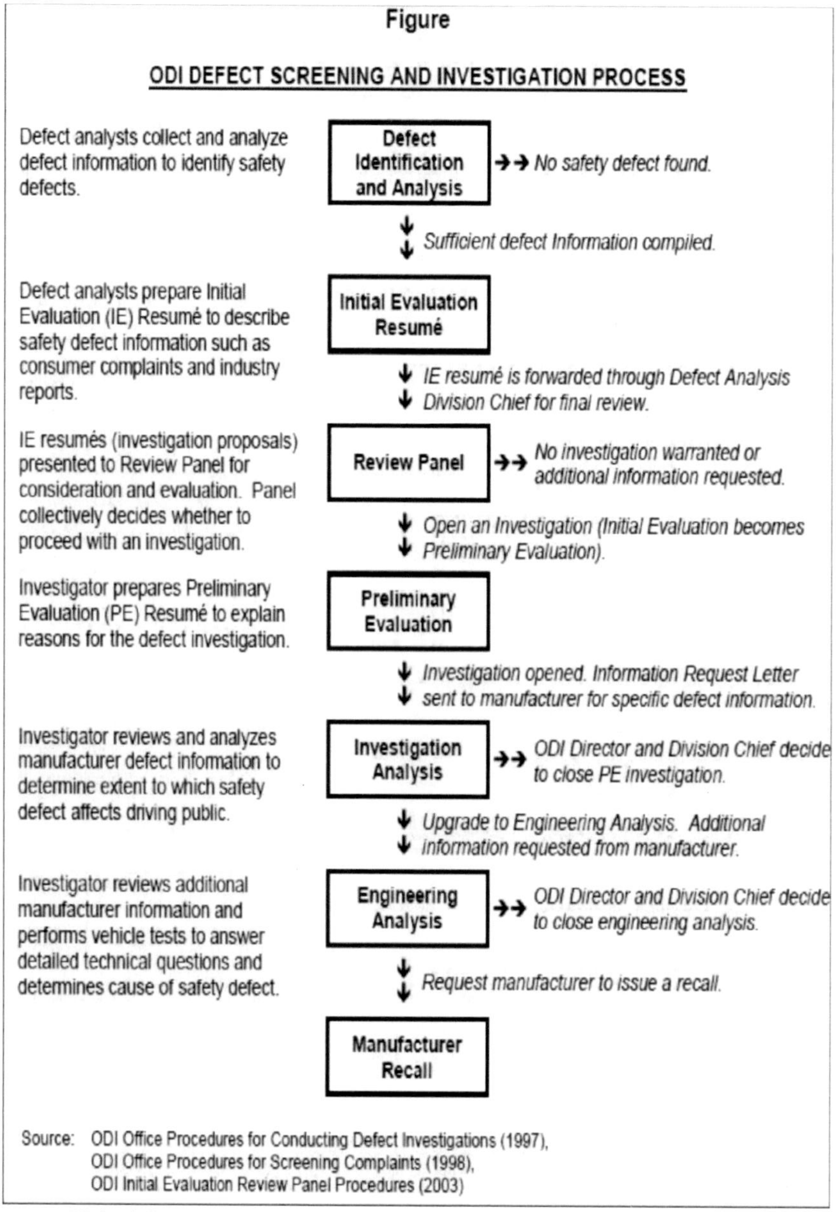


Source: United States Department of Transportation, Office of the Inspector General, *Follow-Up Audit of the Office of Defects Investigation*, MH-2004-088, September 23, 2004, p. 38, http://www.oig.dot.gov/sites/dot/files/ pdfdocs/mh2004088.pdf.

Figure B-1. ODI Defects Investigation Process.

One issue raised by the DOTIG review was the limited amount of time staff had to assess consumer complaints. In its 2002 review the DOTIG estimated that, as ODI was receiving an average of over 34,000 complaints each year, the staff responsible for reviewing those complaints had "an average of about 12 minutes per complaint to review the information, search the defect database for similar complaints, related investigations, and recalls, and decide whether to recommend an investigation."[112]

The relatively limited amount of time each screener has to process each complaint, combined with the number of complaints processed each year and the undoubted variety of the information in each complaint, contributes to another challenge identified by the DOTIG that appears to persist. The DOTIG noted in 2002 that the previous version of NHTSA's defects database contained incorrectly recorded information and that the problems identified in complaints in the database did not always reflect all of the information relevant to the potential defect that was included in the complaint.[113]

Edmunds Automotive in early 2010 did an independent review of the sudden acceleration complaints for all manufacturers contained in NHTSA's current complaints database, and noted that many complaints in the database were miscategorized and many were duplicates.[114]

**Table B-1. Average Number of Issues Evaluated and Presented Per ODI Staff Member**

| Calendar Year | Average Number of Issues Per Staff Member |
|---|---|
| 2000 | 15 |
| 2002 | 20 |
| 2003 | 25 |
| 2006 | 28 |
| 2007 | 35 |

Source: United States Department of Transportation, National Highway Traffic Safety Administration, Budget Estimates for FY2004 (p. 1045-1046), FY2005 (p. 1742, VS-14), FY2006 (1380, III-23), FY2011 (p. 294)

NHTSA reports that now all complaints are initially screened and entered into NHTSA's complaints database by a pair of screeners. These two screeners' responsibilities include categorizing the complaints for further review by a group of six screeners with various areas of expertise who review complaints that fall into their specialties. The fact that only two screeners are responsible for the initial handling the 35,000 or more complaints submitted annually may help explain the problems with the quality of the data in the complaints database.

## Starting an Investigation

The NHTSA peer review panel is composed of representatives from the defects analysis and vehicle investigation divisions. The panel meets once every two weeks to consider the IEs, which propose that an investigation should be opened. If the review panel decides to proceed with an investigation, analysts in the appropriate vehicle investigation division conduct a preliminary evaluation (PE). If the panel decides not to proceed to an investigation, the reasons why are recorded. On average, around 100 investigations are opened each year.

## Stages of Investigation

There are two stages to an investigation: the preliminary evaluation and the engineering analysis.

### *Preliminary Evaluation (PE)*

The PE is based on analysis of additional data obtained from the manufacturer about the potential defect, and may include testing, field investigations, and surveys to determine the number of similar complaints. A PE typically takes four months. A PE may be terminated because the manufacturer initiates a voluntary recall while the PE is underway. At the conclusion of a PE the vehicle investigation division chief and the director of ODI may decide to end the investigation if there does not appear to be evidence of a safety defect, or to proceed to the next stage.

### *Engineering Analysis (EA)*

During the EA, further technical information may be requested from a manufacturer, and further testing, additional field investigations, and more surveys may be done to help determine whether a safety defect exists. An EA can take up to one year.

Given the timing of the elements of a defect identification and investigation, the total defect identification and investigation process can take over a year. NHTSA reports that the average completion time for a defect investigation is eight months.[115] That has been the reported average completion time since at least calendar year 2000.[116]

## Recalls

At the conclusion of an EA, the vehicle investigative division chief and the director of ODI may decide to close the investigation, or they may decide that the evidence warrants a safety recall. In that case, NHTSA is to send a letter to a manufacturer requesting that the manufacturer conduct a recall. NHTSA's Defects and Recall Information Analysis Division monitors recalls. However, as noted above, manufacturers may decide to initiate a recall before NHTSA concludes a defect investigation.

As **Table 3** above shows, the average annual number of defect recalls has been increasing over the past 15 years, though the average annual number of items affected[117] has decreased in recent years. NHTSA reports that it influences around one-quarter of defect recalls, though the average annual percentage has declined somewhat. Though NHTSA-influenced recalls represent around a quarter or less of the number of recalls, those recalls involve a disproportionate number of items, representing half or more of items affected by defect recalls. A primary mechanism of influence is the opening of a defect investigation; NHTSA reports that approximately 50% of all opened investigations result in a safety recall or other manufacturer action to correct a safety problem.[118]

# APPENDIX C. TOYOTA SUDDEN ACCELERATION TIMELINE

A timeline of actions related to unintended acceleration in Toyota vehicles.[119]

- September 2000—Toyota recalls Lexus models in the United Kingdom to prevent floor mats from interfering with gas pedals.
- December 2000—NHTSA received 49,000 consumer complaints for all reasons in 2000, of which about 1,200 or 2% may have claimed unintended acceleration. Of that total, Toyota accounted for 4% of unintended acceleration complaints.

- May, 2003—NHTSA is petitioned to examine unintended, sudden acceleration in the 1997-2000 Lexus LS and GS 4000. After a review, NHTSA denies the petition on the basis that it has found no evidence of a safety defect trend.
- February 2004—NHTSA opens review of electronic throttle control malfunction complaints in 2002 and 2003 models of the Lexus ES300. The investigation is closed in July 2004 after NHTSA identified 20 unique vehicles and two crashes that may have been related to a loss of throttle control, but could not find a vehicle-based cause.
- February/March 2004—State Farm Insurance Co. notifies NHTSA of a trend of rising claims relating to sudden acceleration in the 2002 and 2003 Lexus ES300 and Toyota Camrys.
- August 2005—NHTSA begins investigation of possible uncontrollable acceleration in 2002-2005 model year Camry, Solara, and Lexus ES vehicles. The agency closes the investigation in January 2006 without a determination of a defect.
- July 2006—Toyota tells NHTSA it will conduct a recall of 367,000 Highlander, Lexus RX330, and Lexus RX400h vehicles to fix floor mat retaining clips that may prevent the accelerator pedal from returning to idle.
- September 2006—NHTSA opens investigation of 2002-2006 model year Camrys and Solaras following reports of sudden acceleration. The investigation is concluded without a defect finding.
- 2006—Toyota CEO Katsuaki Watanabe apologies to consumers for quality

  problems and the company delays some new models by up to a half year.
- March 2007—NHTSA begins investigation into possible pedal entrapment in 2007 Lexus ES350 based on five complaints alleging three crashes and seven injuries. The agency upgrades the investigation in August 2007 following a fatal crash involving a 2007 Camry. NHTSA in September determines that floor mat entrapment caused the accident and orders a Toyota recall. Toyota recalls 55,000 all-weather floor mats for 2007 and 2008 model year Camry and Lexus ES350 vehicles. NHTSA issues Consumer Advisory warning about the need to properly secure or remove floor mats.

- April 2008—NHTSA begins investigation into unintended acceleration, due to trim interference, in the 2004 model Sienna minivan. In January 2009 Toyota recalls 26,501 Sienna minivans to fix the carpet.
- August 2009—California Highway Patrolman Mark Saylor and family members are killed in a Santee, Calif. crash of a rental 2009 Lexus ES350. Investigators determine that pedal entrapment by a floor mat may have contributed to the crash and also raise concerns about accelerator pedal design.
- September 2009—NHTSA tells Toyota it expects a recall to address possible defects in pedal design.
- October 2009—Toyota recalls 3.8 million vehicles in the United States to address the possibility that floor mats could trap accelerator pedals in an open position. The recall later is expanded to more than 5 million vehicles in January 2010.
- November 4, 2009—NHTSA issues press release to correct what it terms inaccurate and misleading information by Toyota regarding the recall of the 3.8 million vehicles. Toyota apologizes for having stated that NHTSA had found that "no defect exists."
- December 15, 2009—NHTSA officials fly to Japan to meet with top Toyota executives to underscore Toyota's obligations under U.S. law and to prod for faster action on safety issues.
- January 16, 2010—Toyota informs NHTSA that accelerator pedals made by supplier CTS Corp may have a "sticking" defect.
- January 21, 2010—Toyota announces it will recall about 2.3 million Toyota vehicles to fix sticky pedals.
- January 25, 2010—NHTSA tells Toyota it must stop selling vehicles that have acknowledged defects, even if it does not have a remedy.
- January 26, 2010—Toyota announces it will suspend sales and production of eights of its models due to concerns regarding sticking accelerator pedals. The Toyota announcement included the 2009-2010 RAV4, 2009-2010 Corolla, 2009-2010 Matrix, 2005-2010 Avalon, 2010 Highlander, 2007-2010 Tundra, 2008- 2010 Sequoia, and some 2007-2010 Camrys. The move did not affect Lexus or Scion vehicles. The decision to cease production affected Toyota plants in Canada, Indiana, Kentucky, and Texas.
- January 29, 2010—NHTSA opens investigation into CTS pedals.
- February 2, 2010—NHTSA renews investigation into Toyota's electronic throttle system.

- February 4, 2010—NHTSA opens investigation into at least 124 consumer complaints about brakes on Toyota Prius hybrids.
- February 8, 2010—Toyota announced a voluntary recall of about 133,000 model year 2010 Prius vehicles, as well as 14,500 Lexus vehicles in order to update software in the auto braking systems. The company said some owners of the Toyota models in question had reported "experiencing inconsistent brake feel during slow and steady application of the brakes on rough or slick road surfaces." The company also says it had made changes in the Prius production.
- February 22, 2010—Toyota says it received a federal grand jury subpoena from the Southern District of New York for documents related to unintended acceleration of its vehicles and the Prius braking system. It also disclosed a similar document request from the Securities and Exchange Commission.
- March 2, 2010—Transportation Secretary Ray LaHood tells the Senate Commerce Committee that the Obama Administration is considering requiring brake override systems on all vehicles with electronic throttle controls.

## End Notes

[1] Ray LaHood, Secretary of Transportation, Testimony to the Senate Committee on Commerce, Science and Transportation, March 2, 2010, http://commerce.

[2] Toyota Motor Sales, "Customer FAQS Regarding the Sticking Accelerator Pedal and Floor Mat Pedal Entrapment Recalls," Press Release, February 3, 2010, http://pressroom.toyota.com/pr/tms/customer-faqs-regarding-the-sticking153495.aspx. Toyota initially recalled 3.8 million vehicles in October, 2009 for issues relating to brake pedal entrapment, but expanded the recall to another 1.1 million vehicles in January 2010. Toyota on January 21 recalled 2.3 million vehicles to repair potentially sticky accelerator pedals. A number of cars were recalled for both pedal entrapment and sticky pedal issues.

[3] Toyota Motor Sales, "Toyota Temporarily Suspends Sales of Selected Vehicles," Press Release, January 26, 2010, http://pressroom.toyota.com/pr/tms/toyota/toyota-temporarily-suspends-sales

[4] U.S. Department of Transportation, "U.S. Transportation Secretary Ray LaHood Announces Major Investigations to Resolve Issue of Sudden Acceleration," Press Release, March 30, 2010.

[5] P.L. 106-414; H.Rept. 106-954, p. 7.

[6] Ibid.

[7] Department of Transportation, "Secretary LaHood Announces DOT is Seeking Maximum Civil Penalty from Toyota," Press Release, April 5, 2010, http://www.nhtsa.gov/.

[8] Although there has been some criticism of NHTSA for not responding sooner, NHTSA initiated eight examinations, including three formal investigations, of sudden unintended acceleration in Toyotas between 2003 and 2009.

[9] Yoshimi Inaba, President and COO Toyota North America and Chairman and CEO of Toyota Motor Sales, testimony to the Senate Committee on Commerce, Science and Transportation, March 2, 2010, http://commerce.senate.gov/ public/?a=Files.Serve&File_id=55 1444b7-8cd8-4999-8735-ec2943 1aa3fe.

[10] Joan Claybrook, President Emeritus, Public Citizen, testimony before the U.S. Congress, House Committee on Oversight and Government Reform, *Hearing on Toyota Gas Pedals: Is the Public at Risk?*, February 24, 2010.

[11] Ray LaHood, Secretary of Transportation, Testimony to the House Committee on Oversight and Government Reform, February 24, 2010, http://oversight TESTIMONY-LaHood.pdf.

[12] United States Department of Transportation, Office of Inspector General, *Review of the Office of Defects Investigation*, MH-2002-071, January 3, 2002, p. 1, http://wwwoig.dot. gov/sites/ dot/files/pdfdocs/mh 200207 1. pdf.

[13] Greiling Keane, Angela and Jeff Plungis, "Toyota Acceleration Tied to 43 Fatal Crashes," *Bloomberg*, March 2, 2010.

[14] David Strickland, NHTSA Administrator, Testimony to the House Energy and Commerce Subcommittee on Commerce, Trade and Consumer Protection, March 11, 2010, http://energycommerce.house.gov/Press_111/ 20100311/ Strickland.Testimony.pdf.

[15] United States Department of Transportation, National Highway Traffic Safety Administration, *FY 2011 Congressional Budget Justification*, p. 55, http://www.dot.gov/budget/2011/ budgetestimates/nhtsa.pdf.

[16] NHTSA, "Toyota Sudden Acceleration, Background & Timeline of NHTSA Actions," Fact Sheet.

[17] Ray LaHood, Secretary of Transportation, Testimony to the Senate Committee on Commerce, Science and Transportation, March 2, 2010, p. 7.

[18] Safety Research and Strategies, Inc., "Sudden Unintended Acceleration Redux: The Unresolved Issue," *The Safety Record*, v. 6, 13, June-July 2009, http://www.safetyresearch. net/2009/07/20/sudden-acceleration/.

[19] These studies include John Pollard and E. Donald Sussman, *An Examination of Sudden Acceleration*, NHTSA, 1989, DOT-HS-8007-367; Transport Canada, *Investigation of Sudden Acceleration Incidents*, December 1988; Japanese Ministry of Transport, *An Investigation on Sudden Starting and/or Acceleration of Vehicles with Automatic Transmission*, April 1989.

[20] Clarence Ditlow, Executive Director, Center for Auto Safety, testimony to the Senate Commerce, Science and Transportation Committee, March 2, 2010, http://commerce 8a96-4f79-b6ac-84a6a0e805f5.

[21] Hearit, Keith Michael, and Jeffrey L. Courtright, "A Social Constructionist Approach to Crisis Management: Allegations of Sudden Acceleration in the Audi 5000," *Communications Studies*, vol. 54, Issue 1, Spring 2003.

[22] Huber, Peter, *Galileo 's Revenge: Junk Science in the Courtroom*, Perseus Book Group, 1991, p. 57.

[23] White, Joseph B. and Dionne Searcey, "Audi Case Set Template for Toyota's Troubles," *Wall Street Journal*, March 12, 2010, http://online.wsj.com/article/SB10001424052748704349304575115952186305536.html.

[24] Hearit, Keith Michael and Jeffrey L. Courtright, "A Social Constructionist Approach to Crisis Management: Allegations of Sudden Acceleration in the Audi 5000," *Communications Studies*, vol. 54, Issue 1, Spring 2003.

[25] Pollard, John and E. Donald Sussman, *An Examination of Sudden Acceleration,* Transportation Systems Center, Cambridge, MA,: NHTSA, 1989, DOT-HS-807-367.

[26] U.S. Department of Transportation, National Highway Traffic Safety Administration, "Denial of Motor Defect Petition," DP03-003, Federal Register, vol. 68. No. 183, September 22, 2003, p. 55076.

[27] Ibid. According to NHTSA, there is no set benchmark figure of complaints per 100,000 that indicates problems with a vehicle. The rate of consumer complaints bears on both the question of whether there is a defect and whether the defect is related to motor vehicle safety. Under case law, in assessing whether there is a defect, relevant considerations include the failure rate of the components in question, failure rates of comparable components, and the importance of the component to the safe operation of a vehicle. NHTSA uses a risk analysis approach when examining consumer complaints, weighing for potential safety risks and frequency. In some cases it is sufficient to look at complaint rates, according to NHTSA. In other instances it is important to look at rates normalized by the number of vehicles or miles driven. In other instances, such as early in the life of a vehicle where the consequences of a problem could be deadly, but there may not be sufficient numbers of incidents to determine a rate, the issue could, nonetheless, merit opening an investigation. Further, a defect that results in a loss of vehicle control presumptively presents an unreasonable risk of a crash. But to complete a recall, NHTSA would have to prove that the loss of control was related to a defect, and could not simply rely on complaints alleging a loss of control.

[28] NHTSA, "Toyota Sudden Acceleration Timeline."

[29] Keane, Angela Greiling and Jeff Plungis, "Toyota Sudden Acceleration Tied to 43 Fatal Crashes," *Bloomberg.com*, March 2, 2010.

[30] *Edmunds AutoObserver*, "Question Congress' Hearings Won't Answer: Are Cars Safe?" February 25, 2010, http://www.autoobserver.com/2010/02/question-congress-hearings-wont-answer-are-cars-safe.html. The Big 6 manufacturers are Chrysler, Ford, GM, Honda, Nissan, and Toyota.

[31] "State Farm UIA Claims (Pre-ETC v. Post ETC), NTHSA Document Produced to [House Energy and Commerce] Committee on February 19, 2010," http://commerce 8c5c-4384bd7522ac.

[32] Limbach, James, "Toyota, Lexus Lead the Pack for Sudden Acceleration Problems," *ConsumerAffairs.com*, December 9, 2009, http://www.consumeraffairs.com/news 04/2009/12/cr_sudden_acceleration.html.

[33] Associated Press, "Toyota Will Replace Pedals, If Asked," *Boston Globe*, March 24, 2010, ttp://www.boston.com/business/articles/2010/03/24/toyota_will_replace_pedals_if_asked/.

[34] Office of the Inspector General, United States Department of Transportation, "Audit Initiated of NHTSA's Office of Defects Investigation," February 19, 2010, http://www.oig.dot.gov/library-item/5271.

[35] Ray LaHood, Secretary of Transportation, Testimony to the House Committee on Oversight and Government Reform, February 24, 2010. Department of Transportation, "Secretary LaHood Announces DOT is Seeking Maximum Civil Penalty from Toyota," Press Release, April 5, 2010, http://www.nhtsa.gov/.

[36] Toyota Motor Corporation, "Toyota Motor Corporation Agrees to Settle NTHSA Civil Penalty," Press Release, April 19, 2010, http://pressroom.toyota.com/pr/tms/toyota-motor-corporation-agrees-157093.aspx.

[37] Toyota, *History of Toyota*, Timeline, http://www2.toyota.co.jp/en/history/.

[38] Wards Automotive Group, *U.S. Vehicle Sales Market Share by Company, 1961-2009*, http://wardsauto.com/keydata/.

[39] Bunkley, Nick, "Toyota Ahead of G.M. in 2008 Sales," *New York Times*, January 21, 2009, http://www.nytimes.com/2009/01/22/business/22auto.html.

[40] Oliver Wyman Harbour Report, "Lean Improvements, Worker Buyouts Bring Detroit Three Productivity Closer to Asian Rivals, Says Oliver Wyman's Harbour Report 2008," Press Release, June 5, 2009, http://www.oliverwyman.com/content_images

[41] Center for Automotive Research, *Contribution of Toyota Motor North America to the Economies of Sixteen States and the United States in 2006*, October 2007, p. iii, http://www.cargroup.org/documents/Toyota_16_State _Final.pdf.

[42] Toyota, *Toyota in the United States*, http://www.toyota.com/about/our_business/our_numbers/.

[43] *Automotive News Data Center*.

[44] Toyota Motor Corporation, *FY 2010 Third Quarter Financial Results*, February 4, 2010, p. 20. http://www.toyota.co.jp/en/ir/financial_results/2010/q3/presentation.pdf.

[45] Toyota Financial Services, *Company Background*, http://www.toyotafinancial.com/consumer/tfs.portal?_nfpb=true& _pageLabel=pg_CompanyBackground.

[46] Press, Jim, President of Toyota North America, *A New Era for Toyota and TMA in North America*, September 20, 2006, http://commerce

[47] James Lentz, President and Chief Operating Officer, Toyota Motor Sales USA, testimony to the House Committee on Energy and Commerce, February 23, 2010, http://energy commerce.house.gov/Press_111/20100223/ Lentz.Testimony.pdf.

[48] Akio Toyoda, President, Toyota Motor Corporation, testimony to the House Committee on Oversight and Government Reform, February 24, 2010, http://oversight2010/022410_Toyota/ TESTIMONY-Toyoda.pdf.

[49] Yoshimi Inaba, President and COO Toyota North American and Chairman and CEO of Toyota Motor Sales, Testimony to the Senate Committee on Commerce, Science and Transportation, March 2, 2010, http://commerce.= 3&YearDisplay=2010.

[50] NHTSA and the Insurance Institute for Highway Safety, "Q&As: Event Data Recorders," November 2009, http://www.iihs.org/research/qanda/edr.html.

[51] National Highway Traffic Safety Administration, Department of Transportation, "Event Data Recorders; Final Rule," 71 *Federal Register* 50998, August 28, 2006.

[52] Toyota Motor Sales, 2010 Toyota Electronic Throttle Control Webinar, February 22, 2010, http://pressroom.toyota.com/pr/tms/2010-toyota-electronic-throttle-154266.aspx.

[53] Reuters, "Factbox—What has Toyota pledged to change?" February 23, 2010, http://www.reuters.com/article/ idUSN23 12046420100224.

[54] Toyota Motor Sales, "Toyota Announces March Sales Event," Press Release, March 2, 2010, http://pressroom.toyota.com/pr/tms/toyota-announces-march-sales

[55] Toyota Motor Sales, "Toyota Announces March and First Quarter Sales," Press Release, April 1, 2010, http://www.toyota.com/about/news/corporate/2010/04/01-1-sales

[56] Sanchanta, Mariko and Yoshio Takahashi, "Toyota Recall Losses Could Top $5 Billion," *Wall Street Journal*, March 9, 2010, http://online.wsj.com/article/SB10001424052748704145904575111341893725992.html?K EYWORDS= toyota.

[57] Levinthal, Dave, "Toyota's Lobbying Power Primed for Test as Congressional Scrutiny Mounts," *OpenSecrets.org*, Center for Responsive Politics, February 8, 2010, http://www.opensecrets.org/news/2010 /02/toyotas-lobbying-

[58] Ibid.

[59] Honda, *Operations Overview*, http://corporate.honda.com/america/overview.aspx.

[60] *Bloomberg*, "Regulators Hired by Toyota Helped Halt Acceleration Probes," February 13, 2010, http://www.bloomberg.com/apps/news?pid=20601087&sid=atXvi2msqPOM.

[61] Joan Claybrook, President Emeritus, Public Citizen, testimony before the U.S. Congress, House Committee on Oversight and Government Reform, *Hearing on Toyota Gas Pedals: Is the Public at Risk?*, February 24, 2010.

[62] NHTSA's operations & research budget (i.e., its budget for activities other than safety grants to states) for the period FY1976-1989 peaked, in real terms, in FY1981, so the defects investigation budget for that year was used as the basis for comparison. The 1981 figure was taken from NHTSA's FY1982 budget estimate, and was converted to 2010 constant dollars using the GDP chained price index figures in Table 10.1—Gross Domestic Product and Deflators Used in the Historical Tables: 1940-2015, in Office of Management and Budget, Budget of *the United States Government FY2011: Historical Tables*, pp. 210-211.

[63] FY2000 staffing: Office of the Inspector General, Department of Transportation, *Review of the Office of Defects Investigation*, MH-2002-071, January 3, 2002, p. 34.

[64] House Committee on Oversight and Government Reform hearing, "*Toyota Gas Pedals: Is The Public at Risk,*" February 24, 2010.

[65] To demonstrate that a suspected safety defect exists in a vehicle, an investigation need only find an instance of that defect. It is impossible to prove that a suspected defect does not exist in a vehicle, since to do so an investigation would have to explore every possibility under which that defect might appear. In a machine as complex as a car, the number of combinations of variables that would have to be explored is vast, and the circumstances under which a defect might appear could be extremely rare; consequently, investigations could continue interminably.

[66] United States Department of Transportation, Office of Inspector General, *Review of the Office of Defects Investigation*, MH-2002-071, January 3, 2002, p. 11, http://www.oig. dot.gov/ sites/dot/files/pdfdocs/mh200 2071.pdf.

[67] Jeremy Anwyl, chief executive of Edmunds.com, "No easy answer for the Toyota problem," *Washington Post*, March 16, 2010, p. A19.

[68] John D. Pocari, Deputy Secretary of Transportation, testimony before the Senate Commerce, Science and Transportation Committee, March 4, 2010, http://commerce 2420-4463-a663-72ed299bf55a.

[69] United States Department of Transportation, NHTSA, *FY 2011 Congressional Budget Justification*, p. 54, http://www.dot.gov/budget/2011/budgetestimates/nhtsa.pdf.

[70] U.S. General Accounting Office, *Highway Safety: Research Continues on a Variety of Factors that Contribute to Motor Vehicle Crashes*, GAO-03-436, March 2003, p. 7, http://www.gao.gov/new.items/d03436.pdf.

[71] David Strickland, NHTSA Administrator, Testimony before the U.S. Congress, Senate Committee on Commerce, Science, and Transportation, *Hearing on the Government Response to the Toyota Recalls*, 111th Cong., 2nd sess., March 2, 2010.

[72] Personal communication from NHTSA, March 16, 2010.

[73] Simanaitis, Dennis, "Add More Lightness—Technical Report," *Road & Track*, February 11, 2010, p. 2, http://www.roadandtrack.com/auto_news/tech/add-more-lightness/page_2_-_add_more_lightness_page_2; and "R&T Safety Glossary—Technical Report," *Road & Track*, March 15, 2010, http://www.roadandtrack. com/auto_news/tech/rt-safety

[74] Bates, Hilary, Matthias Holweg, Michael Lewis, Nick Oliver, "Motor Vehicle Recalls: Trends, Patterns, and Emerging Issues," *Omega*, 35 (2) (April 2007), pp. 202-212.

[75] Yi Sun and Chaoqun Jiao, *The Current State of Automotive Electromagnetic Compatibility Research*, Department of Electrical Engineering, Jinging Jiatong University, China., Progress in Electromagnetics Research Symposium, Beijing, China, March 27-29, 2009, http://piers.mit.edu/piers2k9beijing/programfinal.php.

[76] Ibid.

[77] Information from Professor Todd Huber, March 10, 2010.

[78] The Clemson University Vehicular Electronics Laboratory, Automotive Electronic Compatibility (EMC) Standards, CISPR (Automotive Emissions Requirements), http://www.cvel.clemson.edu/auto/auto_emc_ standards. According to the Alliance of Automobile Manufacturers, which represents major U.S. and foreign auto companies, three of the most important electronic guidelines include International Organization for Standardization standards 11451, 11452, and 7637. The European Union also has some mandatory directives.

[79] Toyota Motor Sales, 2010 Toyota Electronic Throttle Control Webinar, February 22, 2010, http://pressroom.toyota.com/pr/tms/2010-toyota-electronic-throttle-154266.aspx.

[80] Broy, Manfred, *Challenges in Automotive Software Engineering*, Institut fur Informatik, Technische Universitat Munchen, 2006. International Conference on Software Engineering, Shanghai, China, May 20-28, 2006, http://www.irisa.fr/lande/lande/icse-proceedings/ cse/p33.pdf.

[81] Ibid.

[82] Ibid. Broy noted data indicating that as many as half of the Electronic Control Units that mechanics replace in cars are technically error free—they are removed because mechanics could not find a better way to fix the problem.

[83] Personal communication with the Alliance of Automobile Manufacturers, April 7, 2010.

[84] Toyota Motor Sales, "Toyota Announces Voluntary Recall on 2010 Model-Year Prius to Update ABS Software," Press Release, February 8, 2010.

[85] Wharton School of Business, "Car Trouble: Should We Recall the U.S. Auto Industry?" Knowledge@ Wharton, May 4, 2005, http://knowledge.wharton.upenn.edu/printer_friendly.cfm?articleid=1183.

[86] Ellims, Michael, Pi Technology, "On Wheels, Nuts and Software," 2004, *Conferences in Research and Practice in Information Technology*, Vol. 47. Paper prepared for 9th Australian Workshop on Safety-Related Programmable Systems.

[87] Ibid. Grimm, Klaus, "Software Technology in an Automotive Company—Major Challenges," 25th International Conference on Software Engineering (ICSE'03), May 2003, p. 498.

[88] Grimm, Klaus, "Software Technology in an Automotive Company—Major Challenges," 25th International Conference on Software Engineering (ISCE'03), May 2003, p. 498.

[89] Hill, Chris, "Making the Most of MISRA C++," *embeddednews.co.uk*, http://www.embeddednews.co.uk/ ArticleItem.aspx?Cont_Title=Making+the+most+of+MISRA.

[90] Tapp, Chris, Keylevel Consultants Ltd., "An Introduction to MISRA," Presentation at University of Warwick, June 2007.

[91] Information from the Society of Automotive Engineers. The guidelines create a SAE Software Assessment Repository, which is a Web-based system that allows organizations that develop embedded automotive software to pose and share data. The repository does not mandate using one assessment other over another.

[92] Autosar, FAQ, http://www.autosar.org/index.php?p=1&up=6&uup=0.

[93] Mauser, Hans and Erwin Thurmer, *Electronic Throttle Control—A Dependability Case Study*, Journal of Universal Computer Science, vo. 5. No. 10, November 28, 1999, http://www.jucs.org/jucs_5_10/electronic_ throttle_control Mauser_H.pdf.

[94] Toyota Motor Sales, 2010 Toyota Electronic Throttle Control Webinar, February 22, 2010, http://pressroom.toyota.com/pr/tms/2010-toyota-electronic-throttle-154266.aspx.

[95] Yoshimi Inaba, President and COO Toyota North America and Chairman and CEO of Toyota Motor Sales, testimony to the Senate Committee on Commerce, Science and Transportation, March 2, 2010.

[96] U.S. Department of Transportation, "U.S. Transportation Secretary Ray LaHood Announces Major Investigations to Resolve Issue of Sudden Acceleration," Press Release, March 30, 2010.

[97] Gilbert, David W., *Toyota Electronic Throttle Control Investigation: Preliminary Report*, February 21, 2010, http://energycommerce.house.gov/Press_111/20100223/Gilbert.Testimony.pdf.

[98] Toyota Motor Sales, "Comprehensive Analysis Raises Concerns about Gilbert Testimony, ABC News Segment," March 8, 2010, Web cast.

[99] Statement of Sen. Rockefeller at March 2, 2010, Senate Commerce, Science and Industry Committee hearing on Toyota recalls.

[100] Testimony of Clarence Ditlow, Executive Director of the Center for Automotive Safety, before the Senate Commerce, Science and Industry Committee, March 2, 2010. Joan Claybrook, President Emeritus, Public Citizen, testimony before the U.S. Congress, House Committee on Oversight and Government Reform, *Hearing on Toyota Gas Pedals: Is the Public at Risk?*, February 24, 2010.

[101] United States Department of Transportation, Office of Inspector General, *Review of the Office of Defects Investigation*, MH-2002-071, January 3, 2002, p. 1, http://www.oig.dot.gov/ sites/dot/files/ pdfdocs/mh200 2071.pdf.

[102] United States Department of Transportation, Office of Inspector General, *Review of the Office of Defects Investigation*, MH-2002-071, January 3, 2002, p. 11, http://www.oig. dot. gov/ sites/dot/files/pdfdocs/mh200 2071.pdf.

[103] United States Department of Transportation, National Highway Traffic Safety Administration, *Public Release of EWR Data*, (no date, but cites a change beginning September 10, 2008), http://www-odi.nhtsa.dot.gov/ewr/qb/ documents/NHTSA-ODI-EWR-Facts.pdf.

[104] United States Department of Transportation, Office of the Inspector General, *Follow-Up Audit of the Office of Defects Investigation*, MH-2004-088, September 23, 2004, pp. 16-17, http://www.oig.dot.gov/sites/dot/files/ pdfdocs/ mh2004088.pdf.

[105] United States Department of Transportation, Office of the Inspector General, *Follow-Up Audit of the Office of Defects Investigation*, MH-2004-088, September 23, 2004, p. 14, http://www.oig.dot.gov/sites/dot/files/ pdfdocs/ mh2004088.pdf.

[106] United States Department of Transportation, Office of the Inspector General, *Follow-Up Audit of the Office of Defects Investigation*, MH-2004-088, September 23, 2004, pp. 16-17, http://www.oig.dot.gov/sites/dot/files/ pdfdocs/ mh2004088.pdf.

[107] United States Department of Transportation, Office of the Inspector General, *Follow-Up Audit of the Office of Defects Investigation*, MH-2004-088, September 23, 2004, p. 14, http://www.oig.dot.gov/sites/dot/files/ pdfdocs/mh2004088.pdf.

[108] United States Department of Transportation, National Highway Traffic Safety Administration, *FY 2011 Congressional Budget Justification*, p. 292, http://www.dot.gov/ budget/ 2011/budgetestimates/nhtsa.pdf.

[109] United States Department of Transportation, Office of Inspector General, *Review of the Office of Defects Investigation*, MH-2002-071, January 3, 2002, pp. 14-16, http://www.oig. dot.gov/sites/dot/files/pdfdocs/ mh2002071.pdf.

[110] United States Department of Transportation, Office of Inspector General, *Review of the Office of Defects Investigation*, MH-2002-071, January 3, 2002, p. 17, http://www.oig.dot.gov/ sites/dot/files/pdfdocs/mh200 2071.pdf.

[111] United States Department of Transportation, National Highway Traffic Safety Administration, *FY 2011 Congressional Budget Justification*, p. 294, http://www.dot.gov/ budget/2011/ budgetestimates/nhtsa.pdf.

[112] United States Department of Transportation, Office of Inspector General, *Review of the Office of Defects Investigation*, MH-2002-071, January 3, 2002, p. 11, http://www.oig.dot.gov/ sites/dot/files/pdfdocs/mh200 2071.pdf.

[113] United States Department of Transportation, Office of Inspector General, *Review of the Office of Defects Investigation*, MH-2002-071, January 3, 2002, p. 11, http://www.oig.dot.gov/ sites/dot/files/pdfdocs/mh200 2071.pdf.

[114] Jeremy Anwyl, Chief Executive, Edmunds.com, "No Easy Answer for the Toyota Problem," *Washington Post*, March 16, 2010, p. A19.

[115] United States Department of Transportation, National Highway Traffic Safety Administration, *FY 2011 Congressional Budget Justification*, p. 55, http://www.dot.gov/ budget/2011/ budgetestimates/nhtsa.pdf.

[116] This eight-month average figure for calendar year 2000 is referenced in NHTSA's FY2004 budget estimate; the figure is repeated in subsequent budget estimates up through the one for FY2011.

[117] "Items affected" includes vehicles, auto equipment (such as tires), and child safety seats.

[118] United States Department of Transportation, National Highway Traffic Safety Administration, *FY 2011 Congressional Budget Justification*, p. 55, http://www.dot.gov/budget/ 2011/ budgetestimates/nhtsa.pdf.

[119] Timeline compiled from *Reuters News Service* and *Automotive News*, "Events Leading to Toyota's Crisis," February 24, 2010; Toyota Corporate Press Releases; congressional testimony and NHTSA documents

In: Unintended or Sudden Acceleration... ISBN: 978-1-61728-983-5
Editor: Denise M. McMann © 2010 Nova Science Publishers, Inc.

*Chapter 2*

# STATEMENT OF THE HONORABLE RAY LAHOOD, SECRETARY OF TRANSPORTATION, BEFORE THE COMMITTEE ON COMMERCE, SCIENCE AND TRANSPORTATION, HEARING ON "TOYOTA'S RECALLS AND THE GOVERNMENT'S RESPONSE"

***Ray Lahood***

Chairman Rockefeller, Ranking Minority Member Hutchison, and Members of the Committee:

Thank you for the opportunity to appear before you today to discuss the important issue of Toyota's recent safety recalls and the broader issue of sudden unintended acceleration. With me today is David Strickland, Administrator of the National Highway Traffic Safety Administration (NHTSA).

Transportation safety is the Department's highest priority. We understand the level of concern about the safety of Toyota vehicles, particularly with regard to unintended acceleration. I would like to explain the recent recalls, the role that NHTSA played in ensuring the recalls occurred, and the actions NHTSA is taking to identify any additional safety defects that might cause unintended acceleration.

The recent Toyota recalls related to unintended acceleration involve two issues: first, accelerator pedal entrapment by floor mats, which can lead to uncontrolled acceleration at very high speeds; and second, accelerator pedals sticking or returning slowly after being depressed, which occurs at a variety of throttle positions but, to the best of our knowledge, is more likely to occur at low throttle positions more readily controlled by the vehicle's brakes.

Before I discuss the details of these two recalls and NHTSA's investigations, I want to clarify what owners of vehicles affected by these recalls should do. To avoid pedal entrapment, remove all floor mats from the driver's side of your vehicle until you receive the repair for this problem from a Toyota dealer. If you do not remove the mat, make sure that it is always securely anchored in place on the retaining hooks and that no other mats are ever stacked on top of it. If your vehicle is covered by the "sticky pedal" recall, pay special attention to your gas pedal. If the pedal is harder to depress or slower to return after releasing it, this could be a precursor to a sticky pedal. If your pedal shows those symptoms you should contact a Toyota dealer immediately. If your accelerator becomes stuck for any reason, steadily apply the brake, put the car in neutral, bring it to a stop in a safe place, and call your dealer.

## Pedal Entrapment

Of the two big recalls, the far more serious problem, in our view, is pedal entrapment by floor mats. We are aware of five deaths that have occurred due to this problem, including a tragedy near San Diego last August that claimed four lives. We have the greatest sympathy for the loved ones of those members of the Saylor and Lastrella families who died in that crash.

Pedal entrapment involves a situation in which the driver intends to accelerate quickly (such as when passing another car or entering a freeway) and depresses the accelerator pedal toward the floor of the vehicle. When pushed far enough the pedal becomes entrapped by the floor mat in full open throttle position. Once the pedal is entrapped, the vehicle will continue to accelerate well in excess of the driver's intent unless the driver can overcome that situation. Given the very high speeds involved and the firmness with which the mat is holding the pedal at full throttle, these are the most dangerous situations we are aware of that come under the broad heading of unintended acceleration. It is very important to note that, even on the recalled vehicles, entrapment by the mat can occur only if the floor mat is out of position because it is not secured, one floor mat is stacked on top of another floor mat, or a floor mat is used that is not intended for use on the vehicle and is inappropriate due to its shape or dimensions.

NHTSA first became aware of this phenomenon in Toyota's Lexus ES350 in 2007 and quickly opened an investigation in March of that year. NHTSA acted based on five complaints from vehicle owners. No related fatalities had been reported at the time the investigation began, but there had been three crashes allegedly related to pedal entrapment by the floor mat. At the time, the problem seemed most likely to occur in Lexus ES350 vehicles where a thick, all-weather floor mat offered as an option by Toyota was used. The shape of these floor mats and a raised portion forming a ridge made them particularly likely to entrap the pedal if not properly secured. So far as NHTSA knew at that time, the accelerator pedals themselves were functioning as designed and the problem centered on the way the pedal could be entrapped by these floor mats under certain conditions.

NHTSA escalated the investigation to an engineering analysis five months later, in August 2007. Shortly before that, a fatal crash involving a Camry occurred that was apparently caused by entrapment. In September 2007, Toyota announced a recall of the all-weather mats in Lexus and Camry vehicles. The remedy was to have the dealers remove the mats and provide a redesigned mat that was shaped in a way that addressed the entrapment risk even if the re-designed mat was improperly anchored.

At the time of the 2007 recall, NHTSA also issued a safety advisory, directed especially to owners of the recalled vehicles but also to all drivers, warning of the serious dangers of not properly anchoring mats or stacking mats on top of each other. At that time NHTSA believed that the recall and removal of the most problematic mats, the improved design of the replacement mats, and education of the public and dealers about the proper use of mats would substantially eliminate the known risk related to pedal entrapment.

NHTSA continued to monitor the situation and became aware of a post-recall crash involving one of the recalled mats that the owner had not removed. Fortunately, that was not a fatal crash but did result in serious injury. In light of that crash and indications that consumer response to this recall was too low, NHTSA urged Toyota to re-notify vehicle owners, which Toyota did in January 2009.

Eight months later, when the San Diego fatal crash occurred on August 28, 2009, NHTSA immediately began to investigate the circumstances of the crash. NHTSA investigators and the San Diego County Sheriff's Department examined the wreckage of the vehicle and concluded that the likely cause was excessive speed due to entrapment of the accelerator pedal by the floor mat. The vehicle was a Toyota Lexus ES350 on loan from a Toyota dealer for the day. The floor mat in the vehicle was designed for a Toyota Lexus RX SUV and was much longer than the mat that would have been proper for the Lexus ES350. At the time NHTSA investigators viewed the wreckage, the accelerator pedal was still fused to the floor mat, apparently melted in that position by the heat of the fire that followed the crash. Combining that observation with the circumstances known to have occurred immediately prior to the crash, including extremely high speeds and the driver's inability to control the speed, NHTSA concluded that the excessive speed was caused by pedal entrapment. Supporting this conclusion was the fact that another customer of the dealership had used the same vehicle just three days earlier and complained of unintended, high-speed acceleration caused by the pedal having been trapped by the mat until he was able to stop the vehicle and free the pedal.

The San Diego tragedy made clear that the entrapment problem could occur in unexpected ways and that recalling the worst performing mats and educating drivers and dealers about not using unsecured, improper, or stacked mats was not going to adequately address the risk. Apparently not even all Toyota dealers were mindful of the need to ensure proper mats and mat anchorage to avoid entrapment.

As a consequence, NHTSA began to explore additional remedial options. The agency continued to review all relevant data to identify any reports that might be linked to similar entrapment in other Toyota vehicles. NHTSA became focused on the pedal design of a number of Toyota vehicles, not because of any known malfunction in their operation but because their shape tended to make entrapment more likely when floor mats are out of position or stacked. NHTSA prepared to open an investigation on the pedal design. At the same time, the agency informed Toyota that the company needed to address this risk promptly as a vehicle defect issue, and requested that Toyota conduct a recall. Toyota responded to NHTSA by announcing a recall to replace or re-shape the pedals in 3.8 million vehicles and sent its official notice of the recall to NHTSA on October 5, 2009.

NHTSA pressed the company to include as part of its recall the addition of a feature called brake override (which some call "smart pedal") technology on models that have keyless ignition systems. With brake override, the vehicle control system gives priority to the signal from the brake pedal and returns the engine to idle when it detects the brake being applied while the accelerator is applied. NHTSA discovered in its investigation of pedal entrapment incidents that in some situations drivers of vehicles with keyless ignition systems did not know that, in Toyota vehicles, they could shut off their engines when in motion only by depressing the dashboard ignition button and holding it for three seconds. The owners were familiar with shutting off the vehicle when it was stopped, which requires holding the button for just one second or less. NHTSA thought it was especially important to ensure that in those vehicles with keyless ignition the driver had the benefit of brake override. Many other manufacturers use this technology and Toyota uses it in newly produced vehicles. The recall Toyota announced in October adhered to NHTSA's request.

NHTSA continued to monitor incoming reports involving relevant incidents. In January, NHTSA told Toyota that its review of other Toyota vehicles indicated that they needed to be included in the pedal entrapment recall. Toyota responded by adding 1.1 million vehicles to the pedal entrapment recall on January 27, 2010.

Under the law, manufacturers have an obligation to notify NHTSA within five days of determining that a defect or noncompliance exists. When manufacturers voluntarily initiate recalls without waiting for NHTSA to order a recall, the process protects the public most quickly. NHTSA can order manufacturers to do recalls but only after initiating a formal investigation, completing its investigation, and following administrative procedures that include a public hearing and opportunities for the manufacturer to file detailed responses. Even after the NHTSA Administrator issues an order directing a recall, the manufacturer can avoid doing the recall until NHTSA proves its case in court. In such a case, the agency has the burden of proving by a preponderance of the evidence that a vehicle defect exists and that it creates an unreasonable risk to safety. As a result, recalls occur most quickly when a manufacturer announces the recall without waiting for NHTSA to open and complete an investigation. That is what happened here—because of the pressure NHTSA applied.

On February 16, NHTSA sent Toyota a Timeliness Query, which is a detailed request for information about when Toyota learned about the defect addressed by this recall. The information Toyota will provide in response to this request will help NHTSA determine whether Toyota's initiation of the recall met its obligation to notify NHTSA quickly. If NHTSA determines that Toyota did not meet that obligation, NHTSA may seek civil penalties from Toyota for that failure. Those penalties could be as high as $16,375,000 for a related series of violations.

## CTS Pedals Sticking

I want to turn now to the "sticky pedal" recall that was initiated in January of this year. NHTSA is not currently aware of any injuries or deaths definitively linked to this problem. Unlike the pedal entrapment recall, which concerns the shape of the pedal that makes it more susceptible to entrapment by an external object (the floor mat), this recall involves the internal working of the pedal assembly. Another distinguishing factor is that the pedal entrapment situations involve instances of full acceleration that are initially intended by the driver, while this problem, to the best of our knowledge, generally involves occurrences at lower power levels where the car continues to accelerate because the pedal does not return upward, or returns slowly, when the driver lessens pressure on the pedal.

The affected pedals are manufactured by CTS Corporation, which is based in Elkhart, Indiana. Some Toyota vehicle owners have complained of certain symptoms in vehicles equipped with those pedals. Those symptoms include a feeling that it is harder than normal to depress the pedal or that, when depressed, it is slower to return. In some circumstances, the situation can involve the pedal not returning at all from the position to which it was depressed. At this time, we understand that this problem is mechanical in nature and does not involve a flaw in the electronic signal being sent from the pedal sensor to the throttle.

In November 2009, NHTSA received several Toyota field reports concerning incidents in which pedals were slow to return or sticking in a number of different Toyota models from various model years. The reports did not indicate a root cause of the symptoms drivers were experiencing. NHTSA reviewed those reports as part of its screening for possible defect trends. Before NHTSA had decided whether or not to open an investigation, Toyota contacted the agency on January 16 about the specific problem it had identified with the CTS pedal. NHTSA told the company it needed a full explanation immediately. Toyota met with NHTSA on January 19 and demonstrated what it thought to be the mechanical problem with the CTS pedals. Based on the information presented by Toyota about the nature of the problem and Toyota's experience with it, NHTSA told the company it expected very prompt action. Two days later, on January 21, Toyota announced the recall, covering some 2.3 million vehicles (many of which are also covered by the pedal entrapment recall and will receive both remedies). Toyota has had the supplier produce a new pedal with a different design that the company believes addresses the issue of excessive friction. The company has also devised an interim remedy to eliminate the safety risk by altering the pedal while new ones are being manufactured. Toyota informed NHTSA that it ceased production of new vehicles in the models affected by this recall so that it could begin to supply the new pedals being produced for the assembly line to dealers for installation in existing vehicles.

On February 16, NHTSA sent Toyota a Timeliness Query about this recall. NHTSA has also begun an investigation to determine whether these particular CTS pedals have been installed in vehicles other than those recalled by Toyota, including those made by other manufacturers. NHTSA will soon receive relevant information from CTS and evaluate it.

## Other Instances of Unintended or Excessive Acceleration

NHTSA receives more than 30,000 complaints from consumers every year concerning perceived safety problems with their vehicles. NHTSA reviews every complaint promptly and, if it appears to contain any evidence related to a safety defect trend, the reviewers begin to track that trend for possible investigation. Among those complaints in recent years have been many allegations of unintended or excessive acceleration on vehicles made by Toyota. Of course, during that same period NHTSA has received thousands of complaints containing such allegations concerning the vehicles made by most major vehicle manufacturers.

The agency has also received several petitions requesting that NHTSA investigate unintended acceleration in various Toyota vehicles. When a member of the public petitions NHTSA to investigate a possible defect, NHTSA examines all information submitted by the petitioner as well as all other information relevant to the particular problem cited by the petitioner. Even where NHTSA denies a defect petition, it does so only after conducting so thorough an examination of the issue that it has effectively done a preliminary investigation. Generally, NHTSA will visit the petitioners, interview them about their experiences, examine their vehicles and vehicle history, drive the vehicles, and search the NHTSA data bases for complaints similar to the experiences petitioners had. In some situations NHTSA will conduct more extensive testing of a vehicle of the same make and model as that of the petitioner.

The information NHTSA has received from consumers concerning unintended or excessive acceleration in vehicles can be divided into general categories that include: engine surging that lasts only a second or two; unintended acceleration from a stopped position or very low speed that results in quick movement over a short distance and sometimes results in crashing into an object; and events that begin at high speeds because the driver intended to accelerate quickly and continue for a sustained period of many seconds or minutes beyond what the driver intended. The possible causes of these events that NHTSA has been able to identify include mechanical problems with the accelerator; obstruction of the accelerator by another object; or human error (pressing the wrong pedal).

NHTSA has carefully reviewed all of the information provided by Toyota consumers in complaints filed with the agency to try to find causes for what they were experiencing. NHTSA also reviews Early Warning Reporting information submitted by the manufacturer and other sources of information, including insurance company submissions. For the high-speed events that last for many seconds or minutes, the only cause NHTSA has been able to establish thus far is entrapment of the pedal by a floor mat. The only exception to this has may have been a recent event in New Jersey that apparently did not involve floor mat entrapment but apparently did involve a stuck CTS pedal. Fortunately, the driver was able to bring the vehicle under control and drive it to a dealership. As discussed, the pedal entrapment issue in the recalled vehicles will presumably be resolved by the recall announced in October. The problem experienced in New Jersey will presumably be addressed by the recall of the CTS pedals announced in January.

NHTSA does not contend that the two recalls will fully resolve all concerns about unintended acceleration in Toyota vehicles. However, with one exception, NHTSA has not been able to establish a vehicle-based cause for unintended acceleration events in Toyota vehicles not covered by those two recalls. The exception was a recall of the model year 2004 Sienna vans in 2009 due to a defective trim panel that could, if loosened during servicing, entrap the accelerator at full throttle. That recall also arose from a NHTSA investigation.

NHTSA initiated a Recall Query on February 16 to ascertain whether Toyota has been completely forthcoming with the agency concerning all possible defects in its vehicles that may be causing unintended acceleration. NHTSA will closely review the documents Toyota submits to determine whether the company has additional information not yet shared with the agency that may cast light on possible defects that cause the problem.

Some consumers and others believe that Toyota's electronic throttle control (ETC) systems, and perhaps such systems in other manufacturers' vehicles, are susceptible to inherent design flaws or electro-magnetic interference (EMI) that can theoretically cause unintended acceleration by resulting in incorrect signals to the engine. These types of electronic systems are commonly used by all major vehicle manufacturers. To date, we have not identified any particular crash or unsafe occurrence that can clearly be attributed to such a flaw or the EMI phenomenon in Toyota's vehicles. NHTSA opened an investigation on Toyota's ETC system in 2004, focused on short duration events, and could not find any safety defects in that system at the time. NHTSA looked at short duration events where no brake application was alleged in this investigation so as to screen out events that could have been caused by driver error, to ensure the agency could find a vehicle-based defect if it existed. In 2008, in wrapping up the floor mat investigation, NHTSA went on to look for additional possible causes of unintended acceleration in the Lexus ES350. That work included some limited electronic and magnetic testing but did not reveal a flaw in the ETC system. Since 1980, NHTSA has conducted 141 investigations on throttle control issues in vehicles made by various manufacturers, some of which involved electronic throttles and some the more traditional mechanical throttle systems.

However, to be absolutely sure that the agency is aware of all potential defects, NHTSA is conducting a review of the general subject of possible design flaws in ETC systems and the possible effects of EMI effects on those systems. We have begun by talking to Toyota and other major manufacturers about the design of their systems and how, through failure modes and effects analysis and other standard techniques, they have taken the possible effects of EMI into account in designing those systems. We have just recently received information about another theory concerning a possible design flaw in the Toyota ETC system. We will explore all relevant information in this examination. To be clear, this is a review of the technological issues, not a defect investigation. However, if any of this activity gives us any reason to believe that a defect may exist in Toyota or other vehicles related to design flaws in or EMI effects on ETC systems, we will open a defect investigation. When we have completed these discussions we will decide whether to conduct any additional research projects that might shed further light on the effectiveness of manufacturers' safety control strategies concerning their ETC systems, including the possible role of EMI effects on various electronic.

## OTHER PENDING TOYOTA INVESTIGATIONS

NHTSA has a total of 44 pending defect investigations concerning various manufacturers and a wide range of issues. Of those, five concern Toyota. One of the Toyota investigations is the Recall Query on sudden acceleration discussed above. Two others have gained wide attention and are summarized here.

NHTSA opened an investigation on February 4, 2010, concerning a braking problem on the model year 2010 Prius. The problem involves a momentary loss of braking when the vehicle hits a pothole, bump, or other uneven surface. NHTSA had received more than 100 complaints about the problem, including four alleged crashes involving two injuries. Five days after NHTSA opened its investigation, on February 9, Toyota announced a recall designed to address this problem. NHTSA will closely monitor its implementation. The recall involves over 148,000 vehicles sold in this country, including the model year 2010 Prius and the 2010 Lexus HS250H. While awaiting an appointment to have their vehicles remedied, owners who experience any braking problems should immediately contact their dealers, and all drivers of these cars should allow extra stopping distance until the problem is fixed.

On February 18, NHTSA opened an investigation concerning approximately 487,000 model year 2009 and 2010 Toyota Corolla and Matrix vehicles. The issue concerns the steering becoming unresponsive or loose at highway speeds. NHTSA had received 168 complaints alleging eight crashes (none fatal) at the time this investigation was opened.

As a final note, I would like to make clear that NHTSA has a very aggressive enforcement program that searches constantly for safety defects and noncompliance with the Federal Motor Vehicle Safety Standards. In just the last three years, NHTSA investigations have resulted in 524 recalls in which 23.5 million vehicles were recalled so that safety problems could be fixed. In addition, several million items of motor vehicle equipment (including imported tires, child seats, and motorcycle helmets) were recalled to correct safety problems.

In summary, NHTSA has acted to ensure Toyota recalls on the issues related to unintended acceleration on which we have had evidence indicating the presence of a vehicle defect, i.e., pedal entrapment and sticky accelerators. We stand ready to ensure prompt action on any additional defects that we have reason to believe are present.

Thank you and I look forward to answering your questions.

In: Unintended or Sudden Acceleration... ISBN: 978-1-61728-983-5
Editor: Denise M. McMann © 2010 Nova Science Publishers, Inc.

*Chapter 3*

# STATEMENT OF CLARENCE M. DITLOW, EXECUTIVE DIRECTOR, CENTER FOR AUTO SAFETY, BEFORE THE COMMITTEE ON COMMERCE, SCIENCE AND TRANSPORTATION, HEARING ON "TOYOTA'S SUDDEN UNINTENDED ACCELERATION"

***Clarence M. Ditlow***

Mr. Chairman and members of the Committee thank you for the opportunity to testify on sudden unintended acceleration in Toyota vehicles and the regulatory response of the National Highway Traffic Safety Administration (NHTSA). The Center for Auto Safety (CAS) is a consumer group founded by Consumers Union and Ralph Nader in 1970 to be a voice for consumers on auto safety.

The Toyota Unintended acceleration crisis which has claimed at least 56 lives was a long time building. Draconian cuts in NHTSA's enforcement budget and staffing, failure to follow up on early research into electronic controls and adopting safety standards based on the research, lax enforcement, flawed research on electronic controls, manufacturers exploiting weaknesses in NHTSA 's regulatory programs, inadequate crash data collection programs and failures to implement the Early Warning Reporting System mandated in the TREAD Act all played significant roles. Even worse for consumers is that more Toyota's remain to be recalled.

**Unrecalled Camrys Lead Deaths**: The only Toyota Camrys being recalled are the 2007-10 model years. The Toyota that leads the known death list in unintended acceleration is the 2005 Camry - there are 5 known crashes with 7 deaths. There are 7 other crashes with 8 deaths in 2002-04 and 06 Camrys not subject to the recall according to public records obtained by the Los Angeles Times. Unrecalled 2002-06 Camrys with electronic throttle control total 12 crashes with 15 deaths compared to 6 crashes with 7 deaths for 2007-10 Camrys. The unrecalled 2002-06 Camrys have twice as many fatal crashes and deaths as the recalled 2007-10 Camrys based on public records of know 2002-10 Camrys linked to unintended acceleration.

| Name | Date | State | Model Year |
|---|---|---|---|
| Barbara Schwarz | September 20, 2007 | Yukon OK | 2005 |
| Anne Ezal | February 25, 2007 | Pismo Beach CA | 2005 |
| Guadalupe Alberto | April 19, 2008 | Flint MI | 2005 |
| Ella Mae & Lon Braswell | June 5, 2005 | Athens GA | 2005 |
| Adegoke & Adeolu Aladegbemi | March 1, 2009 | Marietta GA | 2005 |
| Noriko Uno | August 28, 2009 | Upland CA | 2006 |
| NHTSA Withheld Name | March 14, 2004 | HI | 2002 |
| Juanita Grossman | March 16, 2004 | Evansville IN | 2003 |
| Blossom Malick | March 15, 2004 | Delray Beach FL | 2003 |
| Ethyl Marlene Foster | March 14, 2004 | Phoenix OR | 2004 |
| George & Maureen Yago | January 22, 2004 | Las Vegas NV | 2002 |
| Maria Cafua | September 4, 2003 | Wilmington MA | 2002 |

**NHTSA Investigations:** Beginning in 2001 with the introduction of electronic throttle control (ETC) in 2002 Camry and Lexus ES300, consumer complaints increased by 4-fold in Toyota and Lexus models. In response NHTSA received five defect petitions of which it denied four and granted one. It opened three Preliminary Evaluation (PE) investigations, two of which became Engineering Evaluations. None of these investigations was concluded with a vehicle safety recall. The investigations as a whole show significant weakness in the NHTSA enforcement program which

| Investigation | Year/Make/Model | Outcome |
|---|---|---|
| DP04-003 | 2002-03 Camry, Camry Solara, Lexus ES300 | PE04-021 |
| DP05-002 | 2002-05 Camry, Solara, Lexus ES | Denied |
| DP06-003 | 2002-06 Camry, Solara | Denied |
| DP08-001 | 2004-08 Tacoma | Denied |
| DP09-001 | 2007 Lexus ES350, 2002-03 Lexus ES300 | Denied |
| PE07-016/EA07-010 | 2007-08 Camry, Lexus ES350 | 07E-082 |
| PE08-025/EA08-014 | 2004 Sienna | Safety Improve-ment Campaign |

Toyota exploited to avoid recalls until the tragic crash in San Diego in August 2009 that resulted in 4 deaths in a Lexus driven by an experienced highway patrol officer who was unable to bring the vehicle to a stop. But for the crash being caught on a 911 tape, the recent recalls would not have occurred because the crash would have gone unnoticed like so many before it.

**Early Warning Reporting System Failure:** When the TREAD Act was passed in 2000, Congress required NHTSA to set up an Early Warning Reporting System (EWR, named ARTREMIS by NHTSA) to prevent another Ford-Firestone crisis that led to TREAD. Obviously, it didn't work because we now have a Toyota unintended acceleration crisis. The DOT Inspector General has twice criticized EWR which costs $9.4 million to set up through 2004 and an estimated $11.5 million in operating and maintenance costs from 2005 through 2009. According to the IG:

*Although ARTEMIS became fully operational in July 2004, it does not have the advanced analytical capabilities originally envisioned to help point analysts to potential safety defects. For example, the system cannot automatically notify analysts if consumer-reported complaints and manufacturer-reported warranty claims are both increasing due to vehicle steering problems. According to NHTSA officials, delays in acquiring these capabilities will prevent NHTSA from obtaining full value from the EWR information manufacturers report. While ARTEMIS will automatically point analysts to deaths that manufacturers report so that trends in small numbers of fatalities can be detected, ARTEMIS will not, as currently developed, link*

> *deaths to an alleged defect or identify relationships between the categories of EWR information. In short, ARTEMIS cannot perform more advanced trend and predictive analyses that were originally envisioned as being needed to identify defects warranting investigation. . .*
>
> *[T]he public will have access to only a portion of the EWR information being reported by manufacturers prior to NHTSA formally opening a defect investigation. Since only NHTSA will have access to the majority of the EWR information, it is critical that it establish procedures to ensure congressional concerns expressed in September 2000 about NHTSA's ability to use the data it possessed to spot trends related to failures in Firestone tires have been addressed. Consequently, much will be riding on the ability of NHTSA 's eight analysts, who are responsible for reviewing the large volume of EWR information and drawing conclusions about potential safety defects. This will be especially true until such time as more advanced analytical capabilities are acquired to complement ARTEMIS.*

We don't know whether there are data in EWR on Toyota unintended acceleration and what use NHTSA made of it. Unless a defect investigation in the form of a PE or an EA is opened, the public does not have access to NHTSA's analysis of EWR data. One thing is clear - *NHTSA has opened hundreds of investigations under EWR which are not made public like other defect investigations.* We have gotten access to only one EWR investigation so far – Ford Explorer deaths labeled as DI06-Explorer. While NHTSA may refer to these as inquiries, CAS applies the duck test – if they look like a duck, waddle like a duck and quack like a duck, they are a duck. NHTSA's secretiveness in concealing EWR investigations is unreal but for the fact it used to conceal PE investigations. The agency just doesn't like the public to see what it's doing behind closed doors.

The Center filed a FOIA for all EWR investigatory files and lists of EWR investigations but NHTSA responded by asking us to pay $55,000 in advance. We limited our requests to just lists of EWR investigations to see if any inquiries were made to Toyota that would have given an early inquiry into Toyota acceleration but no response yet. There are only two answers to the EWR Toyota unintended acceleration defect – either (1) EWR worked and gave NHTSA a heads up which NHTSA failed to act on or (2) EWR is a $20 million flop in failing to detect the biggest defect that came down the pike since Firestone tires on Ford Explorers. In order to assess NHTSA performance, EWR investigations must be made public.

**NHTSA Electronics Capability:** Sudden unintended acceleration has always been recognized as a serious safety hazard. Early unintended acceleration recalls involved mechanical failures that were easy to detect and remedy. With the advent of electronic ignition systems and cruise control systems in the late 1970's and early 1980's unintended acceleration complaints without clear mechanical failures began to appear. NHTSA opened more and more unintended acceleration investigation. Some resulted in recalls for electronic control failures. The first two Toyota unintended acceleration recalls were for replacement of the cruise control computer which could cause unintended acceleration on start up.[1]

(1) **1989 Sudden Acceleration Study Led to Invalid Rejection of Toyota Complaints:** As investigations mounted into unintended acceleration in a wide range of vehicles, in January 1989 DOT's Transportation System Center (TSC) conducted a review of unintended acceleration in which it concluded that absent evidence of throttle sticking or cruise control malfunction, driver error must have caused the unintended acceleration.[2] The studies by the Institute for Telecommunications Sciences in 1975 and 1976 and their detailed analytical methods were neither cited nor used. TSC also did not look at electronic throttle control or computer software malfunctions. The vehicles examined in the study were 1983-86 models, none of which had electronic throttle controls or advanced microprocessors systems found in 2002-10 Toyota vehicles.

Based on TSC's finding that brakes could stop a vehicle suddenly accelerating from start up, NHTSA ruled out complaints that the brakes failed or could not stop a unintended acceleration from start up as driver error. A classic example of NHTSA's use of the TSC study is its denial of a defect petition (DP03-003) into unintended acceleration in 1997-00 Lexus LS and GS model which had mechanical accelerator cables:[3]

*"At the conclusion of TSC's effort, comprising thousands of person-hours gathering data, comprehensively testing vehicles including their systems and equipment, interviewing owners and drivers, and inspecting crash scenes and the vehicles involved, a report was released with the following conclusion: "For a unintended acceleration incident in which there is no evidence of throttle sticking or cruise control malfunction, the inescapable conclusion is that these*

*definitely involve the driver inadvertently pressing the accelerator instead of, or in addition to, the brake pedal.''*

In the defect petitions, most consumer complaints were excluded because they were long duration events or where the driver said the brakes could not bring the vehicle to a stop. Not a single defect petition resulted in a recall. The one that was granted (DP04-003) and became an investigation (PE04-021) was closed without a recall after NHTSA excluded most complaints.[4]

**(2) Phantom VRTC EMI Interference Test on 2007 Lexus ES350:** In the most crucial investigation, PE07-016/EA07-010, the agency conducted a test of a 2007 Lexus ES350 to: "Determine whether reported incidents of unintended acceleration were caused by a vehicle system malfunction [electronic controls] or mechanical interference [floor mats]." Later during DP09-001 which the petitioner asked the agency to look at causes of unintended acceleration other than mechanical interference such as electronic controls, the agency used the test report from EA07-010 to deny the petition *without even sending a single information request to Toyota.*

This should have been the definitive test of whether it's floor mats or electronic controls. In DP09-001, NHTSA said: "ODI and VRTC also conducted design reviews and testing to evaluate the possibility of other potential causes of unintended acceleration in the subject vehicles. Some of this work is summarized in the following excerpt from the VRTC test report:

*The Vehicle Research and Test Center obtained a Lexus ES350 for testing. The vehicle was fully instrumented to monitor and acquire data relating to yaw rate, speed, acceleration, deceleration, brake pedal effort, brake line hydraulic pressure, brake pad temperature, engine vacuum, brake booster vacuum, throttle plate position, and accelerator pedal position. Multiple electrical signals were introduced into the electrical system to test the robustness of the electronics against single point failures due to electrical interference. The system proved to have multiple redundancies and showed no vulnerabilities to electrical signal activities. Magnetic fields were introduced in proximity to the throttle body and accelerator pedal potentiometers and did result in an increase in engine revolutions per minute (RPM) of up to approximately 1,000 RPM, similar to a cold-idle engine RPM level. Mechanical interferences at the throttle body caused the engine to shut down.*

Yet when CAS filed a FOIA for the test results and test procedure, NHTSA said it had no test data or any records of test procedure. NHTSA couldn't say what it did, how it did it or what the results were.[5]

**Safety Improvement Campaigns & Equipment Recalls:** To make matters worse, in EA07-010, Toyota agreed to only do an equipment recall of 55,000 all weather floor mats, 07E-082. That was a recall destined to fail. The notification letters to owners did not even require the vehicles be brought in for inspection to see what mats were in the vehicles or how they were secured. The equipment recall saved Toyota $100 million in recall costs according to Toyota's own estimate.

The only other investigation that resulted in an action was PE08-025/EA08-014 which resulted in a Safety Improvement Campaign which is not even recognized under the Motor Vehicle Safety Act. After a private meeting between NHTSA and Toyota including three former NHTSA employees representing Toyota (Erica Jones, Chris Tinto and Chris Santucci),[6] Toyota Vice President Chris Tinto agreed to only a Safety Improvement Campaign as follows:

> *Thank you for taking the time to meet with me and my staff on October 14. Toyota has taken your message seriously and is extending this offer to conduct a field action in order to address the concern raised in EA08-014, an investigation into the Toyota Sienna. . . . Toyota has not determined that the condition at issue in EA08-014 is a 'safety-related defect" within the meaning of the federal vehicle safety laws, and - a summarized below - it continues to believe that no such defect exists.*

How anyone can say unintended acceleration is not a safety defect

The first Safety Improvement Campaign came in 1995 when Chrysler balked at recalling minivans for tailgates that spring open in low impact crashes and killed over 40 people. They are not subject to any sanctions under the Safety Act if they are not carried out. They are not safety recalls and they are not as effective as safety recalls in getting defects remedied. N HTSA defends Safety Improvement Campaigns as the only thing they can get the manufacturer to do because the manufacturers otherwise just say no. This is a self-fulfilling prophecy.

The latest manufacturer to join the "just say no" group is Honda on February 26 which refused to do a safety recall on 2005 Honda Odyssey minvans for tailgate lift struts that fail because NHTSA had let Toyota get away with Safety Improvement Campaign on its minivan. The Honda refusal is all the more troubling because NHTSA had conveyed a rare Safety Panel that approved sending a letter to Honda requesting the company to do a voluntary recall. When Honda just said no, the agency blinked and agreed to the non-statutory recall.

**Toyota Knew & Exploited NHTSA's Regulatory Weaknesses:** From 2001 to the October 2009 floor mat recall (09V-388) generated by the August 2009 San Diego crash, all NHTSA's enforcement effort got was an ineffective equipment recall that saved Toyota $100 million and a Safety Campaign that's not enforceable under the law. Why? First, Toyota knew the investigatory system and exploited it. Only some acceleration complaints were submitted. It knew the agency had limited resources and would agreed to do remedies less than a full vehicle recall because the agency needed to move on to other investigations. Toyota didn't tell the agency about foreign recalls for floor mat interference with the gas pedal that would have caused more emphasis on an earlier vehicle floor mat recall. Toyota requested confidentiality for a wide range of materials that prevented full public scrutiny of the record.

(1) **Lax Enforcement Program:** Toyota was well aware of the fact that from 2004 to 2008, the agency stopped imposing civil penalties for failing to do timely recalls and only imposed $150,000 in penalties since then even though Congress increased the maximum penalty from $800,000 to $15 million inflation adjusted to $16.4 million in the 2000 TREAD Act. In August 2004, NHTSA imposed a $1 million fine, about 7% of the maximum against GM in a W/S wiper recall. In the 1970's NHTSA used to routinely obtain fines from $100,000 to $400,000 which represented up to 50% of the maximum fine instead of 7%.

(2) **Inflated Influenced Recall Statistics:** NHTSA tries to make its recall record look good by referring to 524 recalls involving 23.5 million vehicles obtained as a result of its investigations. These numbers are not what they seem to be. First, 9.3 million came from Ford Cruise Control Deactivation Switch Fire recalls where the agency first launched an investigation in 1998 and got a small recall in 1999. After parked Ford's starting catching on fire in garages and burning houses down, NHTSA belatedly opened more investigation and obtained more recalls. But not until October 2009 did NHTSA obtain the last of the Ford Cruise Control Deactivation Switch Fire recalls, some 11

years after its first investigation. Rather than being a regulatory success, this is a regulatory failure.
The number of recalls is unduly inflated by very small vehicle recalls influenced by a single equipment recall. For example, Dometic made defective refrigerators for recreational vehicles and trailers which resulted in a single equipment recall but 77 vehicle recalls in 2008. Similarly, Ricon made defective wheelchair lifts that resulted in two equipment recalls but nearly 100 vehicle recalls of just a few vehicles each. In each case, the real influenced recall was the equipment recall and the vehicle recall inflated the numbers cited by NHTSA. The 524 recalls should be more less than 300 recalls when the incidental small vehicle recalls are excluded..

(3) **Reduced Budget & Programs:** In 1980, there were 146 million vehicles on the road. Today there are 256 million. In 1980, there 119 people in enforcement, today there are only 57. In 1980, NHTSA had 2 cents per vehicle for enforcement, today it has less than a penny. The agency doesn't have its own test facility and must rent space from Honda in East Liberty OH. Anyway one looks at it, the agency is under funded. In terms of safety, the best way to look at it is motor vehicles are responsible for 95% of the nation's transportation deaths but only 1% of the Transportation budget.

(4) **Crash Investigations:** The National Accident Sampling System (NASS) is another system that could have helped detect Toyota unintended acceleration earlier. The current budget is just over $12 million and investigates only 4,000 crashes per year. This compares with a budget of around $10 million per year in the early 1980s providing about 10,000 cases. The original design would have produced nearly 19,000 cases per year which, at current costs, would require a budget of around $60 million.

Had NASS been operating at its original design size, the agency could have spotted the problem with Firestone tires on Ford Explorers much earlier. The savings in life and limb from that discovery, even a few months earlier, alone would have been sufficient to cover the extra cost of NASS at its full design size. Explorers were introduced in 1990 and the defective Firestone tires were on some of the earliest models. If the excessive Explorer rollovers resulting from failures of Firestone tires could have been spotted by the mid-1990s, it could have saved hundreds of lives and at least one billion dollars for Ford & Firestone.

**Conclusion:** Toyota and NHTSA need to move forward. First and foremost, Toyota needs to install electronic brake override systems in all vehicles with electronic throttle control. Toyota must also agree to releasing all information submitted to NHTSA during the investigations and agreed to conduct a fully public engineering investigation of its electronic controls with independent scientists and engineers with no ties to the auto industry.

NHTSA needs to issue safety standards that:

- A new accelerator standard requiring fail-safe protection that updates the existing 1973 standard, which was written before the advent of electronically controlled accelerators.
- A standard requiring electronic brake override in all automobiles.
- A standard providing electronic magnetic interference protection.
- A standard mandating installation of Event Data Recorders, standard read outs for them and the collection of more information including on rollover crashes.

NHTSA needs to make public all its EWR investigations. Full minutes of all meetings with auto industry officials must be made public to prevent secret deals in all types of investigations. All submissions by manufacturers in investigations must be sworn under penalty of perjury. Elimination of non-statutory recalls such as Safety Improvement Campaigns and regional recalls where only some vehicles in some states get recalled. The whole enforcement program needs to be reinvigorated beginning with assessment of penalties at the top of the scale rather than the bottom. When people are killed by vehicle defects, fines should not be measured in a few dollars, if not a few cents per vehicle.

## End Notes

[1] (86V-132, 90V-040). CAS filed a defect petition (DP86-08) on vehicles recalled in 1990 which was denied as there wasn't a "reasonable possibility" of a recall. More complaints led to PE90-021 and a recall.

[2] " An Examination of Unintended acceleration, " HS-807-367, Jan. 1989 – Main Report, App. A-D

[3] Defect Petition DP03-003 Denial

[4] NHTSA Memo to File by S Yon Restricting Scope of PE04-021 Investigation, March 23, 2004.

[5] CAS Letter to NHTSA Administrator David Strickland - 2/2/10

[6] S. McHenry Memo to EA08-014 File, October 15, 2008.

In: Unintended or Sudden Acceleration... ISBN: 978-1-61728-983-5
Editor: Denise M. McMann © 2010 Nova Science Publishers, Inc.

*Chapter 4*

# TESTIMONY OF TAKESHI UCHIYAMADA, EXECUTIVE VICE-PRESIDENT, TOYOTA MOTOR CORPORATION, BEFORE THE SENATE COMMITTEE ON COMMERCE, SCIENCE AND TRANSPORTATION

***Takeshi Uchiyamada***

Chairman Rockefeller, Ranking Member Hutchison, members of the Committee, thank you for inviting me to address you today. My name is Takeshi Uchiyamada. I am an Executive Vice President of the Toyota Motor Corporation, and I am the Chief Engineer for the global company.

Since I was a child, I have been interested in technology and science. Stories about great inventors such as Edison, Bell and Ford fascinated me, and I dreamed of developing a car that everyone around the world would love. From the time I joined Toyota, I have been engaged in developing vehicles and engineering technology with my wonderful and experienced colleagues.

I was fortunate to be the chief engineer of the first generation Prius. I helped plan and develop the first mass-produced hybrid in the world, and this hybrid led other automakers to realize the importance of environmentally friendly technology. What impressed me most is the fact that consumers had greater environmental awareness than we did as automakers. Our customers helped make hybrid cars popular and used widely around the world today.

Today, I would like to focus my comments on Toyota's approach to safety, our views on engine throttle control systems – or ETCS – and how we are applying advanced technology to further address the issue of unintended acceleration.

As Toyota's President Akio Toyoda testified to Congress last week, Toyota's priority has traditionally been the following: First; Safety, Second; Quality, Third; Volume. Our goal in developing safety-related technology is not only to comply with regulations and standards, and to strive for good safety ratings, but also to improve consumer safety in the real world. While concerns have been raised about our electronic throttle control system, this system – used by all major automakers – actually represents a great safety advancement, enabling superior traction control and electronic stability control, among other things.

Because the ETCS controls the engine throttle system, Toyota places the greatest importance on ensuring that the reliability of this system is absolute by undertaking rigorous design and testing processes. Three things ensure this absolute reliability. The first is the fail-safe mechanisms we build into the design. Second is its tolerance to extreme environmental conditions. And third is its resistance to software problems.

The fail-safe systems in Toyota's ETCS are robust. Our design includes two separate central processors – a main central processing unit, or "CPU", and a sub CPU. The two CPUs are both inside the engine control module and they both get the same throttle-related inputs in parallel from the engine sensor network.

The main, or "control" CPU calculates and executes the operating commands for all engine systems. The sub CPU monitors throttle control inputs, throttle control outputs, and main CPU processes. A "watch dog signal" passes between the two CPUs many times per second to confirm that the processors are working correctly. If the two CPUs are not in agreement, or either the main or sub CPU does not receive the "watch dog signal", the engine management system will alert the driver and go into a fail-safe mode operation.

The ETCS is also designed and tested to make sure it withstands all of the foreseeable environments in terms of temperature, moisture, vibration, and electromagnetic interference (EMI). We have testing data that confirms its reliability from all the markets in which we operate worldwide. On EMI, there is no regulation in the U.S., but we test the ETCS to withstand double the European regulation for EMI. In none of these cases has the ETCS failed.

In addition, we test the software in this system extensively both in the design phase and after it is developed to ensure that there is no possibility of "sudden unintended acceleration."

I want to be absolutely clear: As a result of our extensive testing, we do not believe sudden unintended acceleration because of a defect in our ETCS has ever happened. However, will continue to search for any event in which such a failure could occur.

In order to further validate the safety of our ETCS, we have asked Exponent, a world-class engineering and scientific consulting firm, to conduct its own independent, comprehensive evaluation.

We are also addressing the issue of unintended acceleration through new technologies, including event data recorders and brake override systems.

In conclusion, our Prius has changed the global auto industry with its environmental performance. Now, we will strive to continue to be the leader in the area of safety. I will help drive our team's efforts to meet this challenge, ensure our drivers' safety and regain their trust and confidence.

Thank you.

In: Unintended or Sudden Acceleration... ISBN: 978-1-61728-983-5
Editor: Denise M. McMann © 2010 Nova Science Publishers, Inc.

*Chapter 5*

# TESTIMONY OF KEVIN HAGGERTY, BEFORE THE COMMITTEE ON OVERSIGHT AND GOVERNMENT REFORM, FEBRUARY 24, 2010

*Kevin Haggerty*

February 24,2010
Committee on Oversight and Government Reform
2157 Rayburn House Office Building
Washington DC 2051 5-6143

My name is Kevin Haggerty and I owned a 2007 Toyota Avalon. In the past 6 months I have experienced 5 events where my car accelerated on its own. The first few times I experienced the car accelerating without my foot on the gas pedal I was driving through town. The car would go back to its normal rpms after driving a few miles, or after the car was stopped, turned off and restarted. After experiencing the sudden acceleration the third time, I took my vehicle to be checked by my local auto shop. They could not find anything wrong with my vehicle. After 2 more incidents I brought my car to a Toyota Dealership on November 11th, 2009 to be checked. After keeping my car for 2 days they found no unintended acceleration problems and confirmed that the factory mats were installed properly.

Then on December 28th, 2009 I was driving to work on Route 78 in New Jersey. The car began to accelerate without my foot on the gas pedal. As I pushed on the brake, the car continued to accelerate. At that time I was not able to stop my vehicle by pressing on the brake pedal. The only way I was able to slow the car down was to put the car into neutral. I got off the next exit which was the exit for the dealership. Determined to get the car to the dealership and show them first hand that this was happening, I drove approximately 5 miles by alternating from neutral to drive and pressing very firmly on the brakes. On my way there I called them and asked for the service manager to meet me outside. As I pulled into the front of the dealership, I put the car into neutral and exited the car. With the brakes smoking from the excessive braking and the car's rpm's racing, the manager entered my car. He confirmed that the gas pedal was not obstructed, the mats were properly in place and that the rpm's were very high. They contacted a Toyota Tech to come to the dealership and look at my car. He arrived within a few hours.

The dealership had my car for 1 and 1/2 weeks. When I was told the car was ready to be picked up, I asked what problem they had found. I was told by the service manager that "per Toyota" they replaced the throttle body and accelerator assembly including 1 or 2 of the sensors. Since they could not tell me exactly what problem they found with these parts and why they were replaced, I started doing some research about Toyotas online. I came across Sean Kane's name in multiple articles I read and decided to contact him. When I reached him, I explained my situation and expressed my fear of driving this car in light of what just happened. I no longer felt safe in it since nobody could explain why the acceleration problem occurred in the first place. Sean did not have an answer for the cause and was surprised that the dealership replaced parts and witnessed it first hand.

I was then contacted by ABC News and they were interested in doing a follow up story on accelerator problems. ABC also confirmed with Toyota that the parts taken out of my car were sent to Toyota's Corporate offices to be evaluated. I agreed to an interview mainly because I wanted to help people understand how to safely stop a car by putting it into neutral. I continued driving my car out of necessity but refused to put my children in it.

About 3 weeks ago a local dealership owner, after hearing about my story, made me a generous offer on a new vehicle, as well as offered to pay off the balance of the loan on my Avalon. For my safety, as well as the safety of my family, I took him up on his offer.

Kevin Haggerty

In: Unintended or Sudden Acceleration... ISBN: 978-1-61728-983-5
Editor: Denise M. McMann © 2010 Nova Science Publishers, Inc.

*Chapter 6*

# TESTIMONY OF RHONDA AND EDDIE SMITH, BEFORE THE COMMITTEE ON OVERSIGHT AND GOVERNMENT REFORM, FEBRUARY 24, 2010

***Rhonda and Eddie Smith***

We would like to begin by thanking each of the Honorable Members of this committee for inviting us to testify today regarding the much publicized "sudden unintended acceleration"(SUA) that has been, and is currently being experienced by Toyota drivers not only in the United States, but all over the world. SUA has been the cause of numerous deaths and will continue to be, unless addressed by this committee, Toyota and NHTSA. We truly appreciate this opportunity to share our story now, since we have attempted numerous times to bring this deadly problem to Toyota and NHTSA's attention since October, 2006.

My name is Eddie G. Smith and I reside in Sevierville, TN. I am a senior vice president with Citizens National Bank in Sevierville, TN. This is my wife of 38 years, Rhonda Smith. She is a retired social worker with the State of Tennessee. We have 2 children and 2 grandchildren.

We would like to share our experience with you concerning a SUA incident experienced by my wife on October 12, 2006 in our new Lexus 350 ES. This car had 2728 miles on it when this occurred. The vehicle has a keyless, push button ignition and requires a key fob to be present inside the car in order for it to start. My wife was driving the car and called me on my cell phone at work during most this horrifying experience. Rhonda wrote down this experience October 13, 2006, as we knew this was a potential deadly malfunction that could possibly affect many other Toyota drivers. She is going to read her experience to you, as written by her that day.

"I am writing these words to try and convey some of my feelings of a near death experience, which occurred on October 12, 2006 between approximately 10:50 and 11:00 a.m.

On this Thursday, I had planned on visiting my 85 year old father in Knoxville. I was driving my 2007 Lexus 350 ES from my home in Sevierville down Hwy 66 to I-40 East. Upon entering I-40 I accelerated with everyone else, into the flow of traffic. At this point, I merged over into the second lane, NOT going into passing gear.

It is at this time I lost all control of the acceleration of the vehicle. The car goes into passing gear and the cruise light comes on. At this time, I am thinking that maybe the cruise is what has caused the car to accelerate, as my foot is NOT on the gas pedal. I take off the cruise control. The car continues to accelerate. The car is now up to 80 mph. The brakes do not slow the car at all. Now I am at 85-90 mph. I push the car into NEUTAL and it makes a revving noise. I push the emergency brake on... nothing helps. I continue hitting and slamming the brakes. Now I am at 85-90 mph. I look at the traffic ahead to see if I can maneuver in and out of the upcoming cars and trucks, or if I am going to need to put the car into the guardrail and into the trees.

The last time I looked at the speedometer it read 100 mph. At this time, I had the emergency brake on while frantically shifting between ALL the gears (besides park) but mainly had it in REVERSE and with the emergency brake on. I finally figured the car was going to go to its maximum speed and was praying to God to please help me. After about 3 miles had passed, I thought it was my time to die, and I called my husband (on bluetooth). I knew he couldn't help me in this particular situation, but I just needed to hear his voice. What an awful 911 call he received at work.

At almost exactly 6 miles God intervened. I had not tried anything different that I had frantically tried before to slow the vehicle, yet the car began to slow down ever so slowly. It slowed enough for me to pull to the left median, with the motor still revving up and down. At 35 mph it would not shut off. Finally, at 33 mph I was able to turn the engine off. However, the radio remained on and I was not about to touch ANY button on that car, or ever again."

# RHONDA C. SMITH

Eddie arrived approximately 5-10 minutes later at my location, still on the phone with me. After trying to calm me down he inspected the vehicle to make sure it was turned off and secure. At this time he made a visual inspection for anything out of the ordinary that might have caused this. There was nothing unusual concerning the accelerator or anything that might have jammed it open. The car radio and inside lights were still on, even though I had exited way away from the car and had turned it off. He immediately called a wrecker, as I knew I was never going to drive that car again. After the wrecker arrived, the wrecker driver was given the key fob and asked to tow it back to Sevierville. When he hooked to the car and prepared to winch it onto the rollback, he asked my husband to put the car in neutral so he could start his winch. The driver was standing 20-25 feet away at the rollback controls. Without thinking, my husband sat down in the car with no fob on him and was able to pull the car from park to neutral. This should not have happened. As the car went into neutral, the car actually tried to start by itself with the engine turning over several times. This shocked my husband and he immediately exited the vehicle. At that time the wrecker driver walked back and they were dumbfounded as to how the vehicle could try to start with no key fob inside the vehicle. We have a notarized statement from Tommy Clayton, the driver with Baker's Wrecker Service, stating this fact to be true. The vehicle was then towed back to Sevierville and then to the dealership in Kingsport, TN. The dealership was told this story and they advised they would thoroughly check it out. After several weeks we were advised they could find nothing wrong with the car. We refused to accept this answer and attempted to contact Toyota by phone to let them know that we felt they had an electronics issue that could lead to serious injury and death. Toyota advised they would check on our situation and contact us. After a week to 10 days we had not received a call back. We called again and got the same story of "we will check on it and contact you as soon as possible". This happened the same way several times over the next few weeks and we finally forced a written reply from them that stated, and I quote, "when properly maintained, the brakes will always override the accelerator". Once again we contacted our dealer and expressed our disgust with Toyota's handling of this and they advised us to contact NCDS (National Center for dispute Settlement) and ask for an arbitration hearing. We were eventually set up by NCDS to have a hearing in Gatlinburg, TN, on March 1, 2007. The representative for Lexus was Mr. Leonard St. Amand, their TN District Field Technician. Mr. St. Amand, although in

Kingsport at the time (a 1 hour drive), did not bother to show up but chose to attend via speaker phone. This turned out to be a farce with Mr. St. Amand stating he could not re-create the incident and that we had more than likely caused this problem by standing on the brakes while spinning the tires. Of course we were insulted and furious over being called liars. Needless to say, NCDS denied our claim for a total refund of our money for this "possessed" car, and for the record we did not owe any money on this vehicle.

During this time we had contacted NHTSA and after some prodding, we were contacted by Mr. Steve Chan and Mr. D. Scott Yon, Safety Defects Engineers. This was about the middle of March, 2007.

Mr. Yon took over our claim and seemed to be very receptive of our concerns for this SUA (Sudden Unintended Acceleration) causing serious injury and possibly death. We furnished pictures of the car and documentation of what had transpired since October, 2006. On April 11, 2007, Mr. Yon flew to Knoxville, TN, and drove to Sevierville to inspect our vehicle. The vehicle had been towed to a local Sevierville car dealer's lot and secured for Mr. Yon's inspection. He seemed to arrive with the pre- conceived idea to sell to us, that it was a floor mat problem. We continually insisted that it was not the mats, but instead somewhere in the electronics. Mr. Yon, along with my husband, took the vehicle on a short test drive. Mr. Yon performed several tests at a speed of 50 mph or less. These tests included placing the car in neutral while accelerating, and trying to stop the vehicle with the accelerator engaged and the foot brake fully applied. The transmission did disengage when put in neutral, but the car would not come to a complete stop with the foot brake engaged. Upon returning to the car lot, Mr. Yon and my husband placed the vehicle on a hydraulic lift and removed the wheels and tires. All of the brake pads were totally burnt up and the rotors and drums were ruined. Eventually this was something we had to pay to repair ourselves.

After insisting it was "probably" floor mats, Mr. Yon issued his final report and put the blame on the floor mats. These floor mats were a heavy gauge rubber mat placed on top of the summer mats by the dealer. It would have taken a magic trick for this mat to turn up enough or slide forward enough to cause this SUA. The report was issued on May 2, 2007. In it Mr. Yon claimed to have performed a test with the floor mat in our presence that would show cause for the floor mat to be blamed. This was never demonstrated to us or shown to us that it could ever happen accidently. Once again we advised NHTSA and Mr. Yon that this SUA problem was going to eventually cause the loss of life and serious injury.

At this point we contacted numerous news agencies across the United States, only to have one local station take an interest in our claim of Toyota and NHTSA ignoring this potential deadly problem. Don Dare with WATE-TV, Channel 6 News, Knoxville, TN called and did a full broadcast interview with us, showing where Toyota and NHTSA claimed that SUA was caused by floor mats. He also showed that we believed it to be somewhere in the electronics and not the floor mats. This was aired in the spring of 2007. WATE-TV did another broadcast interview in February, 2010 showing the first interview along with the current interview and showcasing the highway patrolman and his family's horrible 911call shortly before their deaths. The 2010 interview seemed to confirm our 2007 assertion that it was not floor mats, but in the electronics. We did forward this 2010 video to Toyota and NHTSA and received no response.

This was all done by the news station and us to hopefully warn the American public that there was a tremendous and possibly deadly issue with Toyota vehicles, and that Toyota and NHTSA were aware of SUA at least since 2006 but chose to blatantly ignore it.

Eventually in early 2008, due to the traumatizing effect this had taken on Eddie and me, we decided to give up our mission to get Toyota and NHTSA to address this electronics problem in the hopes that they had somehow corrected it. Unfortunately, it took almost 4 years, a mass of injuries and numerous lives being lost for Toyota and NHTSA to confess their sins in this joint dismissal, or cover up, whichever the case may be.

In summary, we would like to inform this committee and the American public that we feel we put forth our best effort in 2006 and 2007 to inform Toyota Motor Company and NHTSA of the potential for SUA to become a deadly issue.

Our hopes were that our efforts might help spare the unnecessary injury and loss of innocent lives. However, we failed miserably, all due to Toyota and NHTSA's uncaring attitude and total disregard for human life.

One would think that Toyota, along with NHTSA's help, would have stepped up and used some of their massive profits to address this now major, deadly problem.

It is our hope that this testimony will in some way help the families of those killed and those that sustained serious injuries from SUA. We also hope they will somehow benefit from the knowledge that we provided critical information to Toyota and NHTSA showing that the problem was not floor mats but in the electronics of their vehicles at least 3 1/2 years ago.

Once again, we would like to thank this committee for taking the time to hear our story and hope that somehow Toyota and NHTSA will be held accountable for choosing the path of not preventing the unnecessary loss of life. We would also like to ask this committee to set a higher standard for NHTSA to be held accountable for in the future.

Thank you,

Rhonda and Eddie Smith

In: Unintended or Sudden Acceleration... ISBN: 978-1-61728-983-5
Editor: Denise M. McMann © 2010 Nova Science Publishers, Inc.

*Chapter 7*

# TESTIMONY OF DAVID W. GILBERT, PHD., PROFESSOR OF AUTOMOTIVE TECHNOLOGY, SOUTHERN ILLINOIS UNIVERSITY CARBONDALE, BEFORE THE COMMITTEE ON ENERGY AND COMMERCE, HEARING ON "TOYOTA SUDDEN UNINTENDED ACCELERATION"

***David W. Gilbert***

Chairman Waxman, Sub-Committee Chairman Stupak and the honorable members of the Committee on Energy and Commerce, thank you for holding this important hearing and allowing me the opportunity to testify before you today.

I have been a technical educator involved with automotive diagnostics and trouble shooting for almost 30 years. I have been witness to many evolutionary changes over that time. When I first began teaching in 1981 at Northeastern Oklahoma A&M College, electronic fuel-injected vehicles were relatively new technology. Over the years, automotive technology has continued to progress from fundamental mechanical systems to more sophisticated electrical and electronic systems. Now, as an automotive technical educator at Southern Illinois University Carbondale (SIUC), I have found electrical diagnostic skills to be supremely important diagnosing and repairing modern vehicles. And, I have spent many hours studying and analyzing new electrical circuits and components. Based on my knowledge of real world failures of components, I purposely duplicate multiple types of electrical problems in donated vehicles for my students to study and diagnose. This provides my students an opportunity to analyze wiring schematics and service information, and actively solve diagnostic problems. SIUC Automotive Technology graduates have found employment in virtually every aspect in the automotive industry. Students graduating from SIUC have the technical skills to work closely with automotive design engineers to ensure reliable vehicle service in real-world situations. I believe the exemplary student placement record, is a result of the academic rigor of the program and the emphasis on technical problem solving.

It stands to reason, that my daily teaching responsibilities would include the application and understanding of electronic throttle control diagnostics. I have the unique perspective in my employment, to research and study multiple vehicles and electronic throttle control system designs. In this preliminary report, my initial findings question the integrity and consistency of Toyota Electronic Control Modules to detect potential electronic throttle control system circuit malfunctions. The absence of a stored diagnostic trouble code in the vehicle's computer is no guarantee that a problem does not exist. I instruct all my automotive students with this fundamental statement: You can have a code with no problem - and a problem with no code.

My curiosity in the Toyota electronic throttle control system began simply with a search for the truth concerning sudden unintended acceleration. I recently purchased a 2010 Toyota Tundra, and with the growing attention in the media to what seemed to be increasing events of sudden unintended acceleration, I made the decision to investigate the foundation of these claims on my own. Based on my working knowledge of electronic throttle controls, I did not expect to the system to be easily fooled without detecting a circuit fault and setting a diagnostic trouble code. It was late one evening when I made a startling discovery; electrical circuit faults could be introduced into the electronic throttle control system without setting a diagnostic trouble code. This discovery opened a window of opportunity within the electronic throttle control system for a potential problem with no code.

Without a diagnostic trouble code set, the vehicle computer will not logically enter into a fail-safe mode of operation. All vehicle manufacturers have recognized the importance for electronic throttle control systems to perform exactly as they intended. Since the vehicle computer will only react to defective sensor inputs outside of the range of programmed limitations if the circuit is not defective; it must be good. Knowing that properly operating electronic throttle control system circuits and components are vital to safe vehicle operation, I proceeded to investigate the problem with more urgency. Because of its important role to accurately convey vehicle driver demands for throttle opening, accelerator pedal sensor voltage inputs need to be confirmable by the vehicles computer as absolutely correct. A complete or partial failure of these electrical circuits, sensors, wiring, or actuators in combination with an absence of fail-safe strategies could potentially result in a runaway engine.

The importance of these issues raised in the electronic throttle control system fail- safe strategies should not be underestimated. Sudden unintended acceleration of a vehicle is a very serious safety concern that should be addressed without delay.

Vehicle manufacturers clearly recognized the important requirement for ETC systems to perform exactly as they intended. A failure of the electrical circuits, sensors, wiring, or actuators could potentially result in a runaway engine. Electronic Throttle Control (ETC) systems needed the added redundancy of certain sensors and electrical circuits to ensure safe and reliable operation. In addition, the ECM's were programmed to detect operational abnormalities or defects in ETC components and their related electrical circuits. The intent was to build an ETC system that would always fail-safe in the event of potential problem.

The purpose of my research study was to contribute to a better understanding of electronic throttle control system malfunctions and the fail-safe detection capabilities of selected vehicles equipped with electronic throttle controls. More specifically, this research examined the fail-safe detection capabilities of electrical circuitry designed to prevent sudden or unintended acceleration of electronic throttle controlled vehicles manufactured by Toyota Motor Co. The Accelerator Pedal Position (APP) sensor was identified in the review of manufacturers' service literature as a significantly important ETC input for all vehicles used in the study. Since vehicle driver demands are electrically conveyed through this high priority sensor, basic testing was focused on the APP sensor, voltages, and associated wiring circuits. A secondary purpose was to identify areas of further research of ETC fail-safe detection capabilities of Toyota Motor Co. vehicles and other vehicle brands. This limited analysis attempted to identify and characterize potential safety concerns of Toyota Motor Co. vehicles, as well as other vehicle manufacturers using electronic throttle control systems.

After completing preliminary tests for Accelerator Pedal Position (APP) sensor signal voltages for the Toyota Electronic Throttle System I examined, it was determined that Electronic Control Module (ECM) malfunction detection strategies were not sufficient to identify all types of fundamental APP sensor and/or circuit malfunctions. Some types of Electronic Throttle Control (ECT) circuit malfunctions were detectable by the ECM, and some were not. Most importantly, the Toyota detection strategies were unable to identify malfunctions of the APP sensor signal inputs to the ECM. APP sensor signal circuits must be undeniably correct to electrically convey the appropriate driver commands to the ECM.

With the two APP sensor signals shorted together through a varying range of resistances, all four Toyota vehicles tested thus far reacted similarly and were unable to detect the purposely induced abnormality. The types of signal faults introduced into the APP circuit should have triggered the vehicles' ECM to illuminate a warning lamp within seconds. The ECM should have then set a Diagnostic Trouble Code (DTC), entered the vehicle fail-safe mode, and reduced engine speed and/or power. When the two APP signal circuits are shorted together, the redundancy of the APP circuit design is effectively nullified and lost. In other words, neither of the shorted APP signal circuits can be verified by the ECM as either; correct or incorrect. The condition then exists for a serious concern for driver safety. In the tested Toyota ETC vehicles, incorrect or corrupted APP sensor signal inputs could potentially result in unwanted engine speeds. Additional research should be done to determine if other vehicle manufacturers may have similar inconsistencies in ETC circuit fault detection.

Using shorted APP signal circuit fault conditions purposely installed on the test vehicles, and with known resistance values that would not set a DTC, vehicle operational behaviors were also noted. It was observed that all test vehicles could be operated without the ECM detecting the induced malfunction. Depending on the resistance value of the APP signal circuit fault, a vehicle may or may not experience noticeable changes in accelerator pedal operational behavior. Observed accelerator pedal operational characteristics included: normal response, sluggish response, and travel with inconsistent engine speeds. It is conceivable that a driver of an ETC vehicle may not notice that an APP sensor and/or circuit malfunction currently exists. Without the aid of an illuminated MIL, a driver could be unaware of electrical problems within the ETC system. In addition, the shorted APP signal circuits were connected momentarily to the sensor's five-volt supply circuit with the vehicle in drive. In all test vehicles, the ECM did not set a DTC and the engine speed increased rapidly to full throttle. This result shows that unusual or sudden unintended acceleration of the vehicle was possible in the ETC test vehicles. It should be noted that in all test vehicle cases, the electronic throttle valve instantaneously moved to wide-open position when the fault was introduced. More research should be done to determine the extent of Toyota ETC vehicles that could be affected by this condition.

In review of the Toyota service information, collected vehicle data, and performance observations; some general assumptions can be drawn from the research completed to date. The inability of the Toyota ECM to detect certain types of short circuit malfunctions could fall back to the basic design of the normal APP signal voltage limitations. The parameters for APP signal short circuit fault detection are apparently too lenient. In the Toyota ETC system, the APP sensor signal voltages rise simultaneously in direct response to accelerator pedal depression. With this design, interconnected signal circuits could be more difficult to identify with a circuit fault detection strategy that uses only threshold voltage limitations.

In this preliminary report, the initial findings question the integrity and consistency of Toyota ECMs to detect potential ETC system circuit malfunctions. The importance of these issues raised in the ETC system fail-safe strategies should not be underestimated. While the small sample of Toyota vehicles cannot be representative of all, these primary findings most certainly warrant further investigation and study. Additional Toyota vehicles of different build years and models should be evaluated for their capabilities of ETC system circuit malfunction detection.

A second recommendation should be a thorough technical investigation and evaluation of ETC fail-safe strategies of Toyota, and possibly other vehicle manufacturers, that experience sudden unintended acceleration that do not appear to be caused by floor- mats or sticking pedals. Priority would be studies of identified vehicles with a high incidence of ETC system related incidences, concerns, or failures involving sudden unintended acceleration.

In: Unintended or Sudden Acceleration... ISBN: 978-1-61728-983-5
Editor: Denise M. McMann © 2010 Nova Science Publishers, Inc.

*Chapter 8*

# TOYOTA ELECTRONIC THROTTLE CONTROL INVESTIGATION: PRELIMINARY REPORT*

***David W. Gilbert and Omar Trinidad***

## INTRODUCTION

Historically, vehicle engine speed was mechanically controlled by use of a linkage or cable to the throttle valve assembly. The introduction of computerized engine controls has given vehicle engineers the added capability to electrically control fuel delivery, ignition, and most recently the throttle valve. By providing electrical throttle control through the means of an on-board vehicle computer, or Electronic Control Module (ECM), a number of engineering advantages were realized. Vehicle manufacturers were able to increase vehicle stability, achieve better fuel economy, reduce emissions, and eliminate components such as cruise control. One of the downside tradeoffs for these beneficial performance advantages, was increased electrical and electronic complexity. Vehicle manufacturers clearly recognized the important requirement for ETC systems to perform exactly as they intended. A failure of the electrical circuits, sensors, wiring, or actuators could potentially result in a runaway engine. Electronic Throttle Control (ETC) systems needed the added

* This is an edited, reformatted and augmented version of a Safety Research and Strategies publication dated February 2010.

redundancy of certain sensors and electrical circuits to ensure safe and reliable operation. In addition, the ECM's were programmed to detect operational abnormalities or defects in ETC components and their related electrical circuits. The intent was to build an ETC system that would always "fail-safe" in the event of potential problem.

Vehicle engineers also needed to build in driver warning capability for ETC systems. In the event that a vital ETC component or electrical circuit defect has been detected, the Malfunction Indicator Lamp (MIL) is illuminated to inform the vehicle operator of a problem. At the same time, to aid in vehicle diagnostics and repair, a Diagnostic Trouble Code (DTC) will have been recorded in the ECM memory. Each possible DTC has an identifying alpha-numeric code to aid vehicle repair of computer associated components. The recorded DTC, along with a "freeze-frame" of captured vehicle data, could be retrieved later by technicians using a diagnostic scan tool. Depending on the seriousness of the ETC malfunction, the ECM will typically have been preprogrammed to operate the engine with limited RPM or substantially reduced power. In the event that mechanical components in the throttle body were to fail, the ECM has the added fail-safe capability to cut-off fuel delivery to limit engine speed. In the most serious cases of component failure, the ECM's fail-safe strategy could have the engine speed reduced to idle. This reduced power mode of operation would remain in the vehicle until proper repairs are made. All of these ETC fail-safe strategies were logically put in place to keep the vehicle safe and driver controllable at all times. However, for fail-safe vehicle operation, the ECM must have the programmed ability to detect a number of potential ETC malfunctions.

The ECM, in addition to operating the engine for optimum performance and emissions, is constantly monitoring sensors and circuits for incorrect or illogical values. Current production vehicles are required to meet standards for the second generation of On-Board Diagnostics (OBD II). An OBD II vehicle runs self-tests or monitors of the ECM circuitry that are typically classified as one or two trip faults. If a sensor or output circuit was to exceed the ECM's programmed threshold values, the computer should set a DTC. However, depending on the component being monitored, a MIL may not illuminate until the ECM sees the component fail a second time. Two trip fault detection logic is used for less definitive and less serious failures that are not easily distinguished as a problem area. One trip fault detection logic is used for more serious or readily identifiable problems. One trip fault monitors will set a DTC and turn on the MIL after only one instance of detection. Typically, one trip faults include electrical component, electrical circuit malfunctions, detected values that could cause engine damage, or potential safety concerns. Properly operating ETC system circuits and components are vital to safe vehicle operation. All DTC's for Toyota's ETC systems are identified in the service literature as "1 trip detection logic" with ECM code setting ability within seconds of circuit fault detection.

## PURPOSE OF THE STUDY

The purpose of this research study was to contribute to a better understanding of electronic throttle control system malfunctions and the fail-safe detection capabilities of selected vehicles equipped with electronic throttle controls. More specifically, this research primarily examined the fail-safe detection capabilities of electrical circuitry designed to prevent sudden or unintended acceleration of electronic throttle controlled vehicles manufactured by Toyota Motor Co. The Accelerator Pedal Position (APP) sensor was identified in the review of manufacturers' service literature as a significantly important ETC input for all vehicles used in the study. Since vehicle driver demands are electrically conveyed through this high priority sensor, basic testing was focused on the APP sensor, voltages, and associated wiring circuits. A secondary purpose was to identify areas of further research of ETC fail-safe detection capabilities of Toyota Motor Co. vehicles and other manufactured brands of vehicles. This limited analysis attempted to identify and characterize potential safety concerns of Toyota Motor Co. vehicles, as well as other vehicle manufacturers using electronic throttle control systems.

## STATEMENT OF THE PROBLEM

What are the malfunction detection and fail-safe capabilities of Toyota vehicles equipped with electronic throttle control systems?

## RESEARCH QUESTIONS

This research study attempts to answer the following questions:

1. Are the malfunction detection strategies of Toyota Electronic Throttle Control systems sufficient to identify fundamental APP sensor and/or circuit malfunctions?
2. Does the possibility exist for Toyota Electronic Throttle Control systems to operate with undetected APP sensor and/or circuit malfunctions?

## ELECTRONIC THROTTLE CONTROL LOGIC

Toyota's TIS service literature served as an obvious first means to understand the complex electrical and mechanical aspects of the ETC system. Claims of vehicle problems without stored DTC's in the ECM, suggested that a possible fail-safe detection condition may have been overlooked in the diagnostics. A thorough review of the available TIS service information for ECT system operation and diagnostics was conducted.

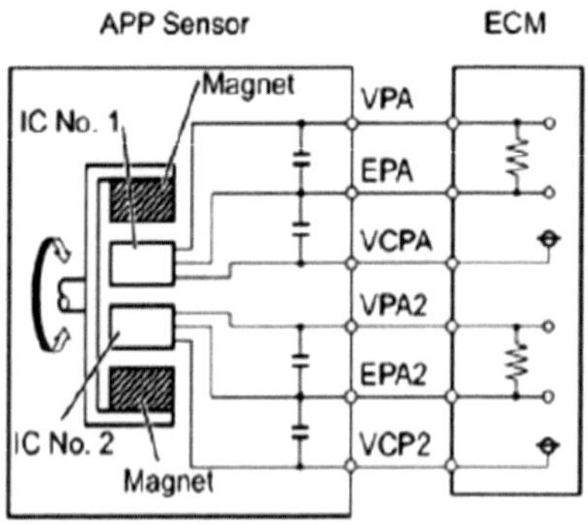


APP Sensor Circuit

Of particular interest were the APP signal inputs to the ECM. According to the Toyota TIS service information, the APP sensor is a non-contact type that uses two independent Hall-effect elements. As described, the APP sensor will have two completely separate sensing circuits: Voltage Pedal Accelerator (VPA), and Voltage Pedal Accelerator 2 (VPA2). Each sensing circuit will have a separate voltage supply circuit (Voltage Constant Pedal, or VCP), a separate ground circuit (Earth Pedal Angle, or EPA), and separate signal circuits (VPA and VPA2). The ETC system description further defines the two signal circuits as VPA (main) and VPA2 (sub). As TIS describes the circuit, the signal from VPA (main) "indicates the actual accelerator pedal opening angle and is used for engine control" (Appendix: TIS Service Document, p. 1). Voltage limits for VPA, range from 0.4 to 4.8 Volts or more. The signal from VPA2 (sub) "conveys the status of the circuit and is used to check the APP sensor itself" (Appendix: TIS Service Document, p. 1). Voltage limits for VPA2, range from 1.2 to 4.8 Volts or more. Of important note, are the *overlapping* voltage values of the two APP signal inputs to the ECM.

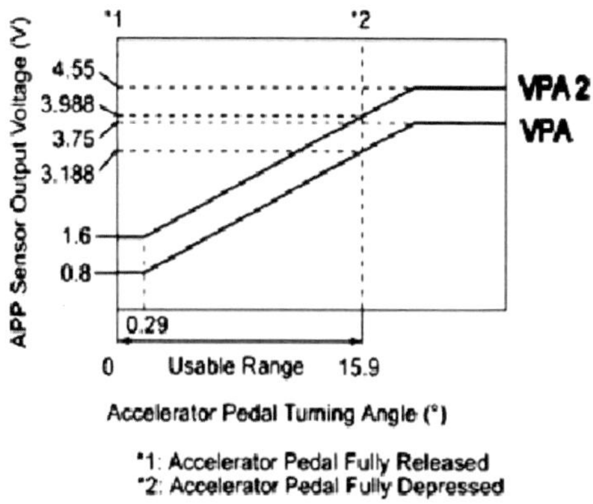


Normal Toyota APP

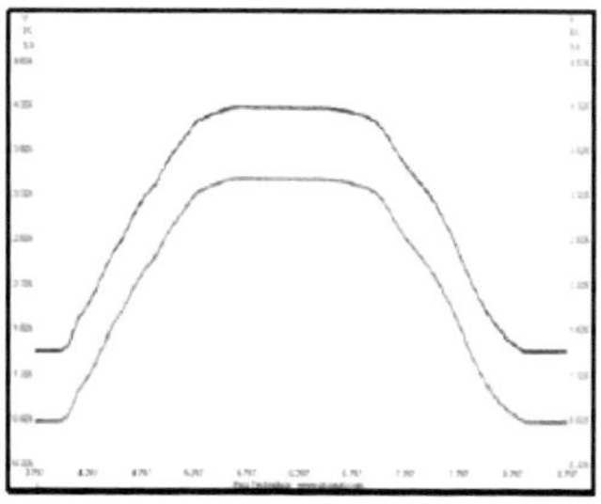

## POTENTIAL LOSS OF APP SENSOR REDUNDANCY

Because their important role to accurately convey vehicle driver demands for throttle opening, APP sensor voltage inputs should always be confirmable by the ECM as absolutely correct. It was noted in the service literature that the threshold limits for a voltage difference of "0.02 Volts or less" between the two APP signal circuits should set a DTC P2138 (Appendix: TIS Document, p. 3). This voltage appeared to be an unusually close threshold value for a short between circuits. In other words, these two signal voltages could have been *almost* identical, and remain safely within the parameters of the P2138 DTC threshold limits. The VPA and VPA2 normally operate through a common range of parallel planes of voltage values. In addition, the two signal voltage values are normally supposed to rise and fall in unison with each other. If the two signal voltages were in some fashion to become interconnected (shorted) through a certain amount of circuit resistance, the lower VPA voltage could be pulled up slightly in value. The higher VPA2 voltage could be be affected in the opposite manner and pulled down slightly in value. At the same time, both sensors could conceivably stay within the upper voltage boundary of 4.8 Volts or more for the P2 123 and P2 128 DTC criteria. Similarly, the VPA2 1.2 Volt low threshold limit value for a P2 127 DTC would not be reached, and the VPA 0.4 Volt low threshold limit value for a P2 122 DTC would not be reached. Signal interconnection through resistance could then potentially tie the two circuits together without setting a DTC. At that point, APP signal circuit redundancy is lost and neither signal circuit is verifiable by the ECM as defective. The ECM will only react to defective voltages outside of the range of programmed limitations-so if the circuit is not defective; it must be good. Without a DTC set, the ECM will not logically enter into a fail-safe mode of operation.

## VEHICLE TESTING METHODOLOGY

The Automotive Technology Department at Southern Illinois University Carbondale (SIUC) has a long history of teaching the technical aspects of vehicles. Working closely with the automotive industry, SIUC has the privileged position of receiving a considerable number of manufacturer donated vehicles for educational and research purposes only. The vehicles used in this research study were purposely selected from the SIUC fleet because of the experimental nature of the testing methodology. As a standard safety and liability requirement of manufacturer vehicle donations, all vehicles used for educational training cannot be licensed, titled, or driven on a public roadway. In addition, when the useful life of the vehicle is exhausted they are destroyed. The SIUC fleet vehicles were clearly a most valuable asset to ensure the safety and control limitations of this unique study.

Test vehicles selected for use in this study were required to meet certain predetermined technical and mechanical requirements. Since ETC was the major emphasis of this investigation, all SIUC vehicles *not* equipped with electronic throttles were logically excluded. Fortunately, the SIUC fleet had four Toyota manufactured ETC vehicles that could be included for testing. Since engine displacement and number of cylinders is not a determining factor in basic ETC operation, vehicles with different engine sizes and configurations were included. Prior to ETC testing procedures, all vehicles used in the study were checked for proper engine performance, electrical system operation, and absence of existing DTC's. Vehicle problems or servicing needs were repaired as necessary before initiating ETC testing procedures.

For each of the test vehicles, ETC specific technical service information was well researched. Collected service information included (but was not limited to) basic ETC operational characteristics, wiring diagrams, electrical connector pin designations, component descriptions, DTC repair procedures, and fail-safe mode criteria. The primary information source for Toyota vehicles was the Technical Information System (TIS), which provides service support for all Toyota produced vehicles in the United States. To ensure technical accuracy before and during testing procedures, hard copies of vehicle service information were printed for laboratory use as needed.

Instruments and test equipment used in this study were typical of those used by automotive service professionals. Tools utilized for circuit testing and evaluation included; Fluke 88 digital multi-meters, Pico Automotive Oscilloscopes, IET decade resistance substituters (variable resistance boxes), connector back-probing devices, and common jumper wires. The high impedance Fluke 88 digital multi-meter was used for testing circuit resistance, grounded circuits, short circuits to power, and shorts between circuits. For purposes of observing and evaluating ETC operational characteristics, a laptop computer based Pico four channel Automotive Oscilloscope was utilized to display and capture waveforms. Simple varying resistance boxes were used to simulate poor circuit connections, short circuits, and mimic defective sensor inputs. Generic OBD II and manufacturer specific scan tools were used to identify the types of DTC's and view the ECM data values for sensor inputs and actuator outputs. Because of availability and versatility to work with many different makes and models of vehicles, the Actron AutoScanner Plus scan tool was used extensively to read DTC's, erase DTC's, and view generic OBD II data during testing.

The first step in the research process was the validation of normal operation of the ETC system in each of the selected test vehicles. Establishment of a solid baseline of proper operation was imperative to evaluate changes in the ETC systems due to the direct effects of testing. As a preconditioning requirement to prevent high idle from cold startup, each engine was started and allowed to reach normal operating temperature before initiating testing methods. General overall engine performance and condition was noted. Scan tools were used to verify the proper operation of the vehicle's MIL and the ECM's capability to detect and record DTCs. Normal scan tool information, scope patterns, and voltage parameters were observed and recorded for all test vehicles used in the study. In addition, general notations were made of the engine speed and response rate with accelerator pedal depression. Engine response was observed during both: accelerator pedal depression with brake pedal released, and accelerator pedal depression with the brake pedal depressed.

Using the vehicle service information as a guide, APP connectors were initially back-probed to tap into the supply, signal, and ground circuits of the sensor. Voltmeter readings were used to verify the circuit integrity of the test connections. Once a solid connection to the circuit was verified, the oscilloscope was connected to visually monitor voltage changes in the APP circuits. Starting from this point, potential types of circuit abnormalities and experimental combinations were examined. Following the fail-safe monitor strategies and the typical malfunction thresholds from the Toyota TIS service literature, DTC detection conditions were painstakingly examined and tested for validity (Appendix: TIS Document).

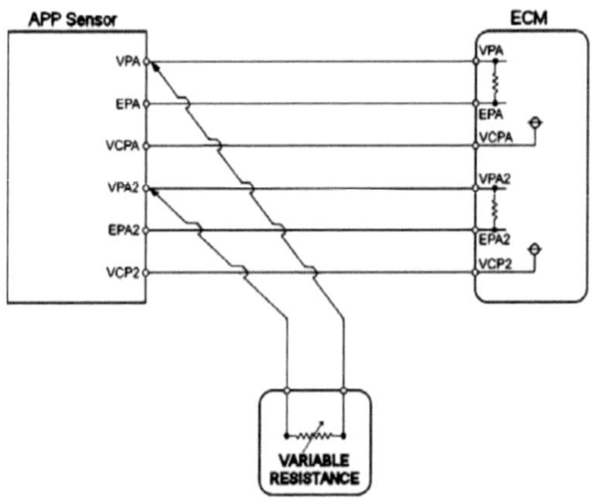


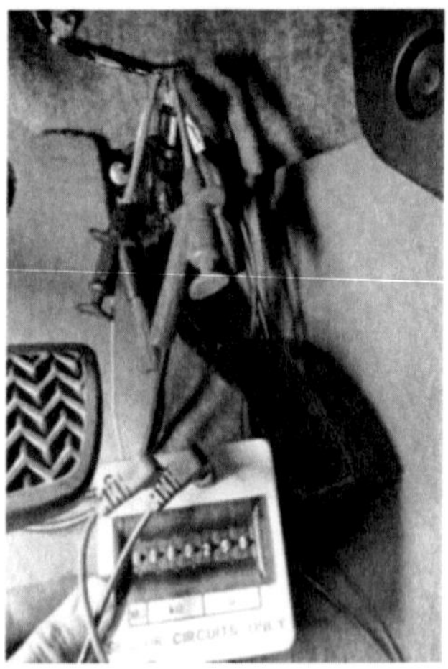

Validation testing began with DTC P2138, which is intended to detect abnormal conditions concerning the APP sensor, the ECM, and short circuits between VPA and VPA2 circuits. While monitoring APP signal voltages with the ignition "Key ON-Engine OFF" (KOEO), various resistances between the two APP signal circuits were introduced for trial. The effects of the different resistance values were recorded to determine the ECM's actual circuit detection capabilities for this type of circuit malfunction. Ample time duration of "2.0 seconds or more" was allowed, as listed in the service literature for one trip detection logic. MIL illumination was used to signal the presence of ECM fault detection and subsequent DTC storage. It was soon realized that certain short circuit resistances between VPA and VPA2 were undetectable by the ECM. Once initial fault limitations were roughly established, "Key ON-Engine Running"(KOER) trials were conducted with similar fault detection procedures. Of particular interest was vehicle engine operation within the ECM acceptable voltage zone of a compromised VPA to VPA2 circuit short. The test vehicles generally ran well, in spite of the simulated signal circuit malfunction.

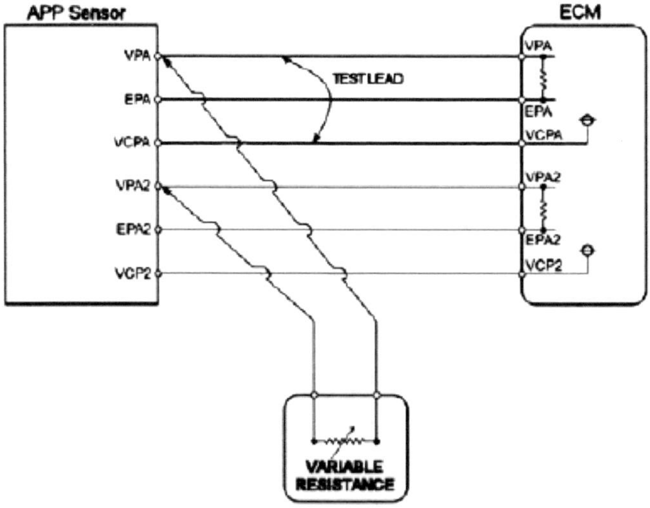


While running the engine in a VPA to VPA2 short circuit state, without MIL illumination or stored DTCs, accelerator pedal response rate and range was noted. Accelerator pedal observations were made with simultaneous brake pedal application, and without simultaneous brake pedal application. Recorded data, observations, and results were listed in detail for each specific test vehicle. To further test the absence of fault detection, experimentation was done to determine if the ECM would set a DTC with a combination of additional circuit abnormalities. Two combination tests were performed to include VPA and VPA2 shorted together (operating undetected), and circuit continuity to ground and power. The VPA and VPA2 signals operating in a short circuit state (undetected by the ECM) were given a circuit path to EPA AND EPA2 (ground). Additionally, the shorted VPA and VPA2 signal circuits (undetected by the ECM) were tested with a circuit path to the VCPA and VCP2 (5 Volt supply). These tests were made during both: KOEO, and KOER conditions. Several different combinations of trials were made before discovering that when VPA and VPA2 were shorted together, with an undetectable resistance, and a connection was made from either VCPA or VCP2 to the VPA2 circuit *only* – both VPA voltages would rise together in unison. As a direct result of the connection to the VCP 5 volt supply circuit, the ECM responded by opening the engine's throttle. Surprisingly, the ECM was repeatedly unable to detect this serious circuit fault abnormality. The vehicle showed no MIL illumination nor stored DTC in the ECM. Most alarming during KOER trials, the engine speed increased almost instantaneously to the full operating RPM limits of the engine. The resulting condition was also present with or without brake pedal application.

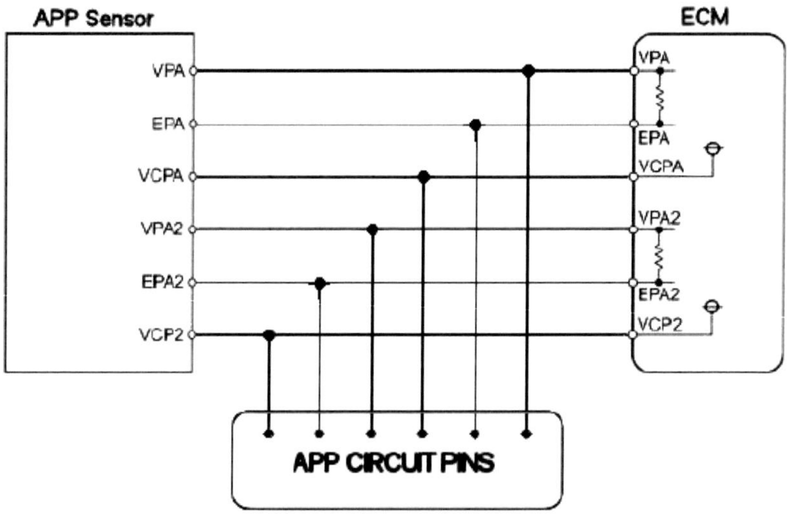


At this point, the back-probing of the APP circuits and ETC testing procedures were re-evaluated. All Toyota test vehicles were considered operationally unstable, with the known potential to reach Wide Open Throttle (WOT) anytime after the engine was started. For more extensive investigation on the research vehicles, wire test leads were tapped directly into the APP wiring harness. The decisive reasoning for installing wire taps was to provide more secure and stable connections to the APP circuitry. By providing more accessible and manageable "pin-out" connections of the APP circuits, testing methods were enhanced with more reliable voltage and resistance measurements. Plus, secure and clearly labeled circuit connections provided for a much safer laboratory testing environment. Evaluation tests were now more narrowly focused on the capabilities of the ECM to detect fault conditions to set a DTC P2138. DTC P2138 is the code used to detect the presence of a short between VPA and VPA2 signal circuits. Resistance values for shorted VPA and VPA2 signal circuits, in combination with a short to the VCP 5 Volt supply circuit, were examined and recorded for threshold parameters. In addition, vehicles were tested to see if these shorted signal circuit conditions and WOT, could be active while the vehicle transmis sion was in drive and service brakes applied and released. For obvious safety reasons, these more extreme tests were performed outside the laboratory in an uninhabited open parking area.

## SUMMARY

After completing preliminary tests for APP sensor signal voltages for the Toyota Electronic Throttle System, it was determined that ECM malfunction detection strategies were not sufficient to identify all types of fundamental APP sensor and/or circuit malfunctions. Some types of ETC circuit malfunctions were detectable by the ECM, and some were not. Most importantly, the Toyota detection strategies were unable to identify malfunctions of the APP sensor signal inputs to the ECM. APP sensor signal circuits must be undeniably correct to electrically convey the appropriate driver commands to the ECM.

With the two APP sensor signals shorted together through a varying range of resistances, all four Toyota vehicles reacted similarly and were unable to detect the purposely induced abnormality. The types of signal faults introduced into the APP circuit should have triggered the vehicles' ECM to illuminate a MIL within seconds. The ECM should have then set a DTC, entered the vehicle "fail-safe" mode, and reduced engine speed and/or power. When the two APP signal circuits are shorted together, the redundancy of the APP circuit design is effectively nullified and lost. In other words, neither of the shorted APP signal circuits can be verified by the ECM as either; correct or incorrect. The condition then exists for a serious concern for driver safety. In the tested Toyota ETC vehicles, incorrect or corrupted APP sensor signal inputs could potentially result in unwanted engine speeds. Additional research should be done to determine if other vehicle manufacturers may have similar inconsistencies in ETC circuit fault detection.

Using shorted APP signal circuit fault conditions purposely installed on the test vehicles, and with known resistance values that would not set a DTC, vehicle operational behaviors were also noted. It was observed that all test vehicles could be operated without the ECM detecting the induced malfunction. Depending on the resistance value of the APP signal circuit fault, a vehicle may or *may not* experience noticeable changes in accelerator pedal operational behavior. Observed accelerator pedal operational characteristics included: normal response, sluggish response, and travel with inconsistent engine speeds. It is conceivable that a driver of an ETC vehicle may not appreciatively notice that an APP sensor and/or circuit malfunction currently exists. Without the aid of an illuminated MIL, a driver could be unaware of electrical problems within the ETC system. In addition, the shorted APP signal circuits were connected momentarily to the sensor's 5 Volt supply circuit with the vehicle in drive. In all test vehicles, the ECM did not set a DTC and the engine speed increased rapidly to full throttle. This result shows that unusual or sudden unintended acceleration of the vehicle was possible in the ETC test vehicles. It should be noted that in all test vehicle cases, the electronic throttle valve *instantaneously* moved to wide-open position when the fault was introduced. More research should be done to determine the extent of Toyota ETC vehicles that could be affected by this condition.

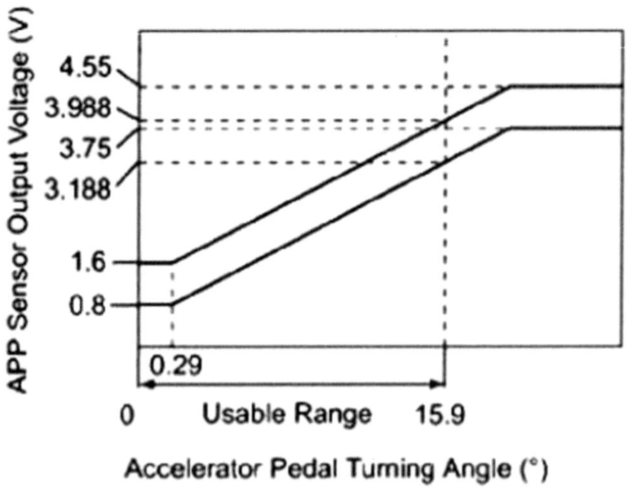


In review of the TIS service information, collected vehicle data, and performance observations; some general assumptions can be drawn from the research completed to date. The inability of the Toyota ECM to detect certain types of short circuit malfunctions could fall back to the basic design of the normal APP signal voltage limitations. The parameters for APP signal short circuit fault detection are apparently too lenient. In the Toyota ETC system, the APP sensor signal voltages rise simultaneously in direct response to accelerator pedal depression. With this design, interconnected signal circuits could be more difficult to identify with a circuit fault detection strategy that uses only threshold voltage limitations.

A more conclusive circuit fault detection strategy could use APP signal voltages (see A & B) that have rising slope, but with offset voltages that increase at slightly different rates. Several vehicle manufacturers currently use this fault detection strategy for APP sensor signal circuits. The obvious advantage, of using two different angles of increasing voltages from the APP signal circuits, is that the signal voltages are never consistently rising at the same rate. In this design, short circuit connections between the two APP signal circuits would ultimately be detected by the ECM because the signal voltages should never increase at the same rate or angle.

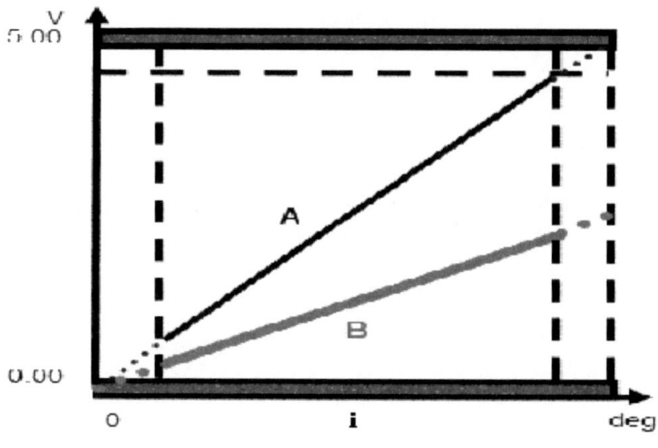


## RECOMMENDATIONS

In this preliminary report, the initial findings question the integrity and consistency of Toyota ECMs to detect potential ETC system circuit malfunctions. The importance of these issues raised in the ETC system fail-safe strategies should not be underestimated. Sudden unintended acceleration of a vehicle is a very serious safety concern that should be addressed without delay. While the small sample of Toyota vehicles cannot be representative of all, these primary findings most certainly warrant further investigation and study. Additional Toyota vehicles of different build years and models should be evaluated for their capabilities of ETC system circuit malfunction detection.

A second recommendation should be a thorough technical investigation and evaluation of ETC fail-safe strategies of Toyota, and possibly other vehicle manufacturers, that experience sudden unintended acceleration that do not appear to be caused by floor-mats or sticking pedals. Priority would be studies of identified vehicles with a high rate of ETC system related incidences, concerns, or failures involving sudden unintended acceleration.

# APPENDICES

**Toyota Electronic Throttle Control Investigation Vehicles**

1) 2005 Toyota Avalon (4T1BK36B55U001024)
2) 2006 Lexus IS350 (JTHBE262762001849)
3) 2007 Toyota Tundra (5TBBV54177S449783)
4) 2009 Toyota Matrix (2T1GE40E19C001003)

# CHAPTER SOURCES

The following chapters have been previously published:

Chapter 1 – This is an edited, reformatted and augmented version of Congressional Research Service publication, Report R41205, dated April 26, 2010.

Chapter 2 – This is an edited, reformatted and augmented version of prepared Statement of the Honorable Ray Lahood Secretary of Transportation before the Committee on commerce, Science and Transportation U.S. Senate dated March 2, 2010.

Chapter 3 – This is an edited, reformatted and augmented version of prepared Statement of Clarence M. Ditlow Executive Director, Center for Auto Safety on Toyota Sudden Unintended Acceleration before the Commerce, Science and Transportation Committee dated March 2, 2010

Chapter 4 – This is an edited, reformatted and augmented version of testimony of Takeshi Uchiyamada Executive vice President, Toyota Motor Corporation, before the Senate Committee on commerce, Science and Transportation U.S. Senate dated March 2, 2010.

Chapter 5 – This is an edited, reformatted and augmented version of Committee on Oversight and Government reform dated February 24, 2010.

Chapter 6 – This is an edited, reformatted and augmented version of Committee on Oversight and Investigations dated February 23, 2010.

Chapter 7 – This is an edited, reformatted and augmented version of Testimony for the Committee on Energy and Commerce, Sub-Committee on Oversight Investigations Toyota Sudden Unintended Acceleration dated February 23, 2010.

Chapter 8 – Toyota Electronic Throttle Control Investigation, Preliminary Report by David W. gilbert, Ph.D. Omar Trinidad Dated February 21 2010.

# INDEX

## G



## H



## I



## J



## K



## L



## M